# Science and Fiction

T0208995

For further volumes:
http://www.springer.com/series/11657

# Science and Fiction – A Springer Series

This collection of entertaining and thought-provoking books will appeal equally to science buffs, scientists and science-fiction fans. It was born out of the recognition that scientific discovery and the creation of plausible fictional scenarios are often two sides of the same coin. Each relies on an understanding of the way the world works, coupled with the imaginative ability to invent new or alternative explanations—and even other worlds. Authored by practicing scientists as well as writers of hard science fiction, these books explore and exploit the borderlands between accepted science and its fictional counterpart. Uncovering mutual influences, promoting fruitful interaction, narrating and analyzing fictional scenarios, together they serve as a reaction vessel for inspired new ideas in science, technology, and beyond.

Whether fiction, fact, or forever undecidable: the Springer Series "Science and Fiction" intends to go where no one has gone before!

Its largely non-technical books take several different approaches. Journey with their authors as they

* Indulge in science speculation—describing intriguing, plausible yet unproven ideas;
* Exploit science fiction for educational purposes and as a means of promoting critical thinking;
* Explore the interplay of science and science fiction – throughout the history of the genre and looking ahead;
* Delve into related topics including, but not limited to: science as a creative process, the limits of science, interplay of literature and knowledge;
* Tell fictional short stories built around well-defined scientific ideas, with a supplement summarizing the science underlying the plot.

Readers can look forward to a broad range of topics, as intriguing as they are important. Here just a few by way of illustration:

* Time travel, superluminal travel, wormholes, teleportation
* Extraterrestrial intelligence and alien civilizations
* Artificial intelligence, planetary brains, the universe as a computer, simulated worlds
* Non-anthropocentric viewpoints
* Synthetic biology, genetic engineering, developing nanotechnologies
* Eco/infrastructure/meteorite-impact disaster scenarios
* Future scenarios, transhumanism, posthumanism, intelligence explosion
* Virtual worlds, cyberspace dramas
* Consciousness and mind manipulation

Dirk Schulze-Makuch

# Alien Encounter

A Scientific Novel

2nd Edition

 Springer

Dirk Schulze-Makuch
School of Earth and Environmental Sciences
Washington State University
Pullman, Washington
USA

Self-published by Dirk Schulze-Makuch, 2008 under the following title "Voids of Eternity: Alien Encounter".

ISSN 2197-1188          ISSN 2197-1196 (electronic)
ISBN 978-3-319-01960-4          ISBN 978-3-319-01961-1 (eBook)
DOI 10.1007/978-3-319-01961-1
Springer Cham Heidelberg New York Dordrecht London

Library of Congress Control Number: 2013946548

*Cover illustration:* Astronaut in the tunnels of the spacecraft. Copyright by iurii /Shutterstock.

Printed on acid-free paper

Springer is part of Springer Science+Business Media (www.springer.com)

*For my wife Joanna and our children—Nikolas, Alexander, Alicia, and Kristian.*

# Preface

Many seem to think that most of the "big discoveries" have already been made. I beg to differ. The same misconception was also very common among physicists before the discovery of relativity theory and quantum physics in the early 20th century. The truth is that we are still far away from understanding the universe and its greatest mystery: life. We have unraveled some working mechanics and details, yet a greater understanding still eludes us. This is particularly true for the phenomenon of life. We do know what happened in the first fractions of a second after the Big Bang, yet we still don't know how life originated on our home planet. Was it only an incredibly unlikely event that occurred only once in the galaxy or universe? Or, is it a common occurrence and has happened on many worlds? And what are the conditions which have to be present for it to happen? Does it always have to happen under the same conditions as for life on Earth—conditions which we don't know and can only speculate about—or are there multiple pathways from inanimate matter to life? Also, once life is there, does it generally stay microbial or will it inevitably become more complex, macroscopic and eventually intelligent? These are some of the open scientific questions that this work addresses in an adventurous setting in which a handful of astronauts explore other worlds and seek to figure out puzzling phenomena.

Perhaps the most intriguing question is the why: why is there a universe and why are we here? While many of my colleagues would argue that this is a religious or spiritual question that cannot be answered by science (which I don't disagree with), it is nevertheless one of the most profound questions keeping us awake at night. I do not pretend to have an answer to this question, but simply weave together my understanding of the science of astrobiology with my take on Eastern philosophy to produce what I believe is a logically consistent scenario, which I hope to be inspirational, informative, and entertaining for you, the reader.

Pullmann WA                                                        Dirk Schulze-Makuch
July 2013

# Acknowledgements

Thank you to all who made this possible. I'm grateful for the suggestions in content and language that I received from my colleagues Harry Boehm, Barbara Fossen, Louis Irwin, Philip Rust, Karen Libey, Don Satterlund, Marina Antonio de Sousa, Eric Shulenberg, Joop Houtkooper, David Darling, my editor Christian Caron and my wife Joanna Schulze-Makuch.

Dirk Schulze-Makuch

# Contents

# Part I

## The Novel

# Alien Encounter

## 1  Venus Orbital Station

### 1.1  Station Emergency

A shrieking alarm, echoing through the hallways, awakened Jack and Vladimir from their deep sleep in the space station. Leaping from their beds, they ran toward the Communication and Computer Systems Module, or CCS unit, which jolted in its orbit around Venus and was the heart of the space station. One would think that it had been built by some kid gone wild with a Mechano set. A spider work of alloy struts linked a cacophony of solar panels, hydrogen tanks, and cylindrical modules. This fragile looking structure contrasted sharply with the yellowish sulfur clouds of Venus below that cloaked so many mysteries begging to be solved. The Venus Orbital Station was built to solve these mysteries.

When Jack and Vladimir reached the CCS, the control units were lit up with flashing red lights. Jack cleared the last remnants of sleep from his eyes. His body was wet from sweat and troublesome pictures emerged from the intense dream he just had, but that was not the time to dwell on it. He had to focus all of his attention on the crisis at hand.

"What the hell is going on?" Vladimir shouted.

They were very aware that they were losing altitude and the space station was heating up fast.

"The orbital thrusters must have quit working." Jack yelled as he sprinted toward the engineering module. He ripped off the cover from the control panels. Sure enough, one of the sub-modules connecting the main panel switch with the thruster control unit was blinking, indicating that it wasn't conducting any current. Jack shouted to Vladimir through the intercom,

"We need the YK5 replacement package for the orbital control unit!"

"I'm on my way!"

Jack could barely hear Vladimir's voice through the shrieking alarm. Time was critical. Vladimir frantically searched for the spare part, found it, and headed back to the CCS module to grab the emergency tool box. To Jack, the seconds seemed like hours. Perspiration covered his forehead.

Suddenly Vladimir was next to Jack. Jack, switching off the thruster power supply, inserted the replacement, and re-started the thrusters. The monotonous humming of the orbital thrusters returned. What a comforting sound! The instruments indicated that they were slowly regaining altitude. Everything was functioning properly again. Everything was okay.

They hurried back to the CCS. Jack sat down at the computer console and submitted their preliminary report to NASA headquarters: *Module YK5, which connects control panel and orbital thruster, was inoperative. Replaced same. Probable cause of failure was lack of proper insulation for air filter, which in turn allowed the caustic Venusian atmosphere to infiltrate and wear down the electrical connectors. Need insulation protection and replacement module ASAP.*

Fifteen minutes later a signal from NASA HQ came in and Fred, one of the controllers responsible for the Venus-Earth connection, appeared onscreen. "You guys are slacking! It took you nearly two and a half minutes to fix the problem and send off the report."

Smart ass. "I guess we stayed up too late last night, watching the yodeling Apple Annies on Global TV!" Vladimir retorted, shaking his head. Oh, how he hated these drills. But he had to admit that they kept their responses sharp, and that they provided some relief from the otherwise rather monotonous work.

Fred concluded his transmission with "Over and out."

Finally, they could settle down. Vladimir was an okay guy, with a presence that commanded respect. One couldn't help but like him. He was a deep thinker, an intellectual, and also built like a truck. His thick crop of hair was combed back, his wide forehead prominent. Jack on the other hand was agile, innovative, and ready to take on anything at the drop of a pin. Curiosity was part of his in-born nature, part of why he was here.

The CCS module was the hub of the station. Ten other modules were attached to it by huge bolts. The low-Venus orbit enabled Jack and Vladimir to research the microbial populations in the lower cloud layer of the atmosphere. However, the sulfuric acid-rich atmosphere slowly ate away at the components of the station systems and orbital thrusters. Jack likened the time spent on fixing the outpost to making 'oil changes' every several million kilometers of orbital travel, analogous in a strange way to what he did on his antique internal combustion engine car back on Earth.

The swirling clouds harbored biochemical enigmas, but after so many months even that idea could become mundane, were it not for the occasional drills like today's. There was also the intellectual give-and-take (usually not quite arguments) between the two men that gave Jack great pleasure. Vladimir liked to discuss deep topics, often inspired by being at the frontier in space. They had to maintain a tight organization: breakfast was an occasion. They made instant coffee, sat down, and removed the wrapping from their breakfast packages.

The images from Jack's dream flooded back into his mind. They were so vivid and colorful. Much too strong for a usual dream, more a vision or nightmare, Jack thought. Earth being bombarded, rattled by explosion after explosion, and he wanting to do something about it, but he was being sucked into some kind of a black hole. And he was wearing a monk's robe! While one explosion hit after another, a woman was screaming for him, a woman who seemed to be close to him. In the background a choir was singing some kind of religious doomsday song. He was clinging at the edge of the black hole, while little flashes lighting up the blue marble of Earth in the distance. With each hit he heard another score of people screaming and moaning. Jack turned his eyes.

"What's going on? You are looking troubled?" Vladimir said while taking a sip from his coffee.

"Oh, some kind of weird dream—very intense." Jack described some of the images to Vladimir.

"Some people from my culture believe that these kinds of dreams are a window to the future. There may be some meaning to it, a premonition." Vladimir suggested.

"Oh, no, it's just some ridiculous stuff. I don't believe in dreams, they are just some spinoff from the unconscious mind. I don't want to talk about it any further." Jack did not like to delve into emotional issues or for that matter into anything that could not be explained by logical scientific reasoning. "Let's continue our conversation from yesterday," Jack said, "We discussed that the Universe started with ripples in space-time. And from there …"

Vladimir interrupted Jack. "So, why were the ripples there in the first place?"

"Well, space is not really nothing. You can have particles and antiparticles emerging out of nothing, a sub-particle zoo, really. And then there is the detection of the Higgs boson …"

"Ah, you only believe that because they detected them at CERN near your home town of Geneva. Only you stubborn Germans believe it! No one else!" Vladimir appeared to be more challenging than usual. Perhaps, it was some kind of fundamental difference between the two, that he, Vladimir, was more open to unconventional ideas, while Jack tended to cling more onto standard scientific explanations.

"Swiss!" Jack insisted, shooting Vladimir a quick glance. It was a minor point, but critical for any Swiss.

"Anyway, my point is that we are too concerned with technicalities. But what's behind it? Why are we here, or why is the Universe here? We still don't have any clue. We are in the second half of the 21st century and still don't know more about THAT than we did at the time of the ancient Greeks!"

Jack rubbed his forehead as was his habit when in deep thought. Then he countered, "We know much more about life, planets, and the Universe itself than we ever knew before. For example, we know what happened a small fraction of a second after the Big Bang, and we know how the four fundamental forces unfolded. We developed M-theory and can use it to explain the multiverse hypothesis."

Jack pushed aside a crate of dehydrated beef stew: the sight of it made his stomach churn, and it wasn't with anticipation. They had to eat this disgusting stuff, day after day because NASA had made a mistake and shipped cases of it instead of the regular variety-pack. It was eat it, or starve. Once this tour was over, he never wanted to lay eyes on dehydrated beef stew again.

"Exactly my point. We are good with descriptions, but cannot answer the question '*why*'. Why are we here?" Vladimir's eyes twinkled with excitement.

The millennia-old question was so broad that Jack wasn't sure how to respond. "Simply because the environmental conditions were right in our Universe, compared to a zillion other universes."

"Did it ever occur to you how strange it is that we know a lot about the first fraction of a second after the Universe was created, but still haven't solved the riddle of the origin of life? We have it right in front of us to study and still there is no real testable or believable hypothesis explaining how the first cell originated."

Jack didn't like being put on the spot. "It's just too complex." His voice was rising. "There are too many chemical possibilities and eventualities. The number of possibilities is astronomically high!"

"Maybe there is simply something wrong with our understanding. Maybe we are missing a crucial link?"

"Oh, come on. You don't want to bring back into play the "vitalism" ideas of the nineteenth century? Or, worse, intelligent design by a supreme being? Or the suggestion that we were seeded by some alien intelligence?"

With a cup of coffee in hand, Vladimir headed to the lab module. The biolaboratory module was the smallest work-space on the station. This space was supposed to house sophisticated instruments and experiments. It never did, because funding was diverted by NASA and RASE (Russian Agency for Space and Exploration) to pay for safety upgrades at the Lunar Deep Space Observation Outpost and to expand the facilities on the Martian Base Station. Nevertheless, a high powered microscope graced the lab bench. Jack and Vladimir considered themselves lucky that they received the expensive artificial gravity boosters, which provided one third of Earth's gravity. Unfortunately, Venus had become a backwater outpost, because Earth perceived it as being too hot and too hostile for humans, and as not having the potential for resources that would make it economically feasible to justify further attention.

Vladimir sensed the agitation in Jack's voice. One had to be careful. He and Jack were the lone occupants of this space station, and if they didn't get along, then what? It was better to be open-minded rather than try to be right all the time. Their getting along was an absolute requirement. Coming from different cultures, clearly they thought differently and viewed things differently. Vladimir was Ukrainian, from the Russian Academy of Sciences. Jack was a Swiss-American, raised in Europe, with German as his first language. He had emigrated to America as a young man. Part of Vladimir's personality, perhaps, stemmed from the fact that the Russians were the first to launch a satellite to study the Earth's upper atmosphere. Vladimir could have beaten his chest about that one, but—politely—he didn't.

"I don't feel strongly about any of these ideas," Vladimir relented. "But all options have to be on the table. There must be a reason why we have made so little progress in researching the origin of life. All I'm saying is that we are missing something. Perhaps we haven't been looking into the right things or in the right places?"

Also realizing the turn the conversation had taken, Jack took a deep breath to calm down. Two men living in such close quarters; two men having to do the jobs that normally required six, on a station designed for six, two men without female companionship; well, they could end up not talking to one another, or worse yet, attempting murder! Even highly trained and rigorously-screened astronauts like themselves could still be bugged by one-another's trivial habits. There were the two of them, always together. You just got tired of being around the same person all the time. Sometimes it was tough to hold one's temper and tough to hold one's tongue. But you did it. It was part of the job.

Vladimir changed the topic, and pointed down at the Venusian surface: "Look, it's clearing up."

"Great, we may get some exceptional views. I'll charge up the camera."

Sometimes, Vladimir Kulik and Jack Kenton felt like they were on the most remote outpost in space, billions of kilometers away from the next human being. Technically, that was not true. They were really not all that far from Earth, and Mars Base Station was even further away from Earth. But the Mars base had six inhabitants and was on the surface of a planet that at least had some similarities to Earth. Here, in orbit around Venus, Earth seemed to be even further away than it did from Mars.

The station was orbiting at the interface of the lower atmosphere and the lower cloud deck. Jack's and Vladimir's primary job involved monitoring the microbial populations floating in the Venusian atmosphere at the lower cloud level, at an altitude of about 50 kms. The possibility of microbial life in the Venusian clouds had been advanced more than a century earlier, and was confirmed forty years ago by a sample-return mission. The two men yearned for more excitement as they monitored the dynamic activities of the microbial population and recorded changes in biomass density, locations, and spatial distribution. There was no lab work to do because it was decided that such work could be done cheaper—and better—on Earth.

Vladimir often commented that he would much rather spend time observing the last polar bears in Siberia. However, living on this frontier in space was a privilege few would ever experience, and it did have its gratifying and delightful moments. Once in a while they were able to peek through the lower atmosphere and enjoy stunning views of the Venusian surface. Today was such a day.

"It's time to take readings from the probes again" Jack said, referring to the soccer-ball size measuring devices that were scattered in the Venusian atmosphere recording biological and atmospheric information every hour.

"All right" responded Vladimir. "I'll get the shuttle ready."

Taking another peek toward the Venusian surface while moving slowly through the access tunnel, Jack guessed that the ride today would be bumpy. Several yellowish streaks and eddies were clearly visible in the atmosphere. Although he was a very skilled and seasoned astronaut, Jack always remained a bit anxious about flying, especially when cramped into small spaces such as a little shuttle. The shuttles were clearly not as reliable as the space agencies wanted you to believe. Jack remembered the last accident when one of these shuttles crashed into West Candor Chasma, a side canyon of Valles Marineris on Mars. He just had started his duty on the Venus Orbital Station.

Vladimir greeted him lightheartedly when he entered the shuttle: "I'm wondering how many bugs burned up today?"

Jack just grumbled as he squeezed into the copilot's seat—Vladimir was the official pilot.

Once both men were seated, Vladimir released the clamps and the shuttle dropped out in free fall toward the Venusian surface. Even after so many gravity drops, Jack still felt an uncomfortable cramping in his abdomen. After a few seconds Vladimir initiated the thrusters and steered toward the first probe. Vladimir took the first reading. "3,200 burned up this week," he said.

Jack plugged the numbers into his computer. "That number appears to be higher than normal. There must have been a microbial enrichment or they may have encountered some unfavorable winds."

"You would think that those guys would have developed some mechanism of buoyancy after 4 billion years of evolution."

"Why?" Jack responded. "The system works perfectly well. Their generation time is one week, and then they fall and burn up in the lower atmosphere after an average lifespan of a month. So, only after four and a half reproduction cycles, on average, they encounter their fiery death. This keeps the gene pool renewed and healthy."

"This is exactly my point." Vladimir insisted. "Why did they not develop a better survival mechanism? Why did they not develop some buoyancy mechanism to enhance survival rates and possibly develop towards something like Earth's eukaryotes, microorganisms that at least have a nucleus? Why didn't they come up with a better UV protection than this sulfuric acid-cyclooocta sulfur coating?"

Jack did not respond. He was tired of engaging in one of these endless and fruitless discussions about evolution. Deep inside he had to agree with Vladimir that science did not seem to be making much progress in understanding life in the universe. After the discovery of life in the atmosphere of Venus and the subsurface of Mars, everyone had been extremely excited. But after a while it was realized that all these life forms were bacteria. The public reaction, of course, was pretty mute and of disinterest, because they were only microbes, bugs. None of the discovered organisms were more sophisticated than a prokaryotic microbe on Earth, a microorganism without a nucleus. Also, their DNA was nearly identical to that of Earth's life. In fact, they appeared to be cousins of terrestrial organisms; bacteria related to the early life forms on Earth. The scientific authorities were still arguing on this matter. But it was becoming increasingly clear that all organisms on Earth, Venus, and Mars came from the same genetic stock at some time, came from the same origin, and only branched off from each other after they had been delivered by asteroid or cometary impacts from one terrestrial planet to the other, early in the Solar System's history.

We humans seem to be as isolated as ever, Jack thought. At least now we know nearly for sure that the actual genesis occurred on Mars. Vladimir suddenly turned and slowed the shuttle, awakening Jack from his thoughts. "We are approaching the next probe."

Vladimir slowed down the shuttle to match the probe and extended the shuttle's robotic arm to link up and download the collected data. They also collected an atmospheric sample for further analysis, and then went on to the next probe. "I bet you look forward to your next assignment. Anything must be more interesting than counting microbes and collecting routine samples" Vladimir said.

"Yes, I put my request in three months ago and should be notified just before the end of my duty here."

"That's less than two weeks!"

"The bureaucrats at home are urging me to take a leave on Earth first, but there is no reason for me to go back."

"There may be a chance that you and Amber could reunite."

"No, she broke up with me for good. It just doesn't work between us. She is too much into money and her career. Even when I wasn't at astronaut training, I hardly ever saw her. How can a loving relationship develop if you are never there?"

"You know that is only half the truth. We discussed it before. You made the choice to become an astronaut rather than settling down into an established life with her."

"I guess, in a way, the only thing we had in common was that both of us lacked the commitment for making our relationship work."

"Don't be so negative. There is always a chance for a re-unification. You only have to try very hard. Maybe give it another shot, there are always possibilities!"

"I don't think so" Jack responded and hesitated for a moment, but then it was just bursting out of him, "Right now, she's going through divorce proceedings."

Vladimir remained silent.

Being out here in space, Jack at least didn't have to deal with this troubling truth, or at least so he thought. It certainly affected him on some very deep level. Vladimir commented quite often that Jack appeared unhappy and grumpy, and several times it appeared to him that Jack showed signs of the onset of depression. But each time, Jack was able to turn it around. The freedom of space certainly helped Jack's emotional well-being.

"Well, I'm looking forward to going back to my family" Vladimir finally continued. "Did I tell you that my friends and family are already planning a big vodka party? But I'm more interested to finally see my four month old son for the first time."

"I'm truly happy for you. I can't even imagine how excited you must be."

The conversation was interrupted when they approached the next probe. They continued to collect data, from a total of twelve probes.

"All done. We got the measurements from all probes within a radius of 500 kms of the orbital station" Vladimir said cheerfully. The shuttle docked back at the station. The big Ukrainian was again in a talkative mood, conversing in good English overlain by his Russian accent. "How about we go down to the surface again and do some more exploring?"

"You know we can do that only if there is a specific reason to do so. Otherwise there will be all this complaining again about cost overruns and the operating budget being too expensive."

Vladimir laughed heartily. "What are they gonna do about it? Fire us?"

Jack smiled. In many ways he liked Vladimir's hip-shot ideas. "We wouldn't be good scientists if we couldn't come up with a reason. Remember those strange albedo changes near Maat Mons, which we still don't understand?" Jack said. He was referring to the variations in reflected light from the largest volcano on Venus. "But I think we should start this tomorrow. I've had enough shuttle squeezing for one day. We have to get the heat protection suits ready anyway, and it is already late in the day."

## 1.2  The Signal from the Surface

The yellowish clouds seemed to extend a bit lower today and the surface was nowhere in sight. Both men suited up, happy with anticipation of possible new discoveries to be made. The surface had been extensively mapped, but having a whole planet at your disposal meant there was always a chance to discover something new and exciting. The shuttle was even more cramped today with all the extra gear Jack crammed into it. After an hour of work and preparation they were finally ready. Vladimir released the clamps and the shuttle dropped nearly vertically down to the Venusian surface, heading toward the Ishtar Terra region. Jack felt his stomach pushing into his lungs. They dove through the cloud layer, accelerating steadily toward the blistering hot planetary surface. After a while, they could finally see the surface: details grew slowly larger, with mountains and canyons taking shape. They were still dropping fast, and Jack looked nervously over at Vladimir.

"The control for the starboard thruster is jammed again," Vladimir complained, looking worried. "I thought I fixed it."

This was definitely no drill. The volcanoes of Ishtar Terra raced toward them at phenomenal speed and Jack wondered whether a stupid little malfunction would have them end up crashing on the flank of one volcano. If so, death would come quickly. They still had a few seconds to prevent that…

"Oh, here you go!" Vladimir exclaimed after some wiggling of controls. He was able to initiate the thrusters and redirected the shuttle to fly at a constant altitude of several hundred meters above the surface. Jack took a deep breath.

"What a ride!" said Vladimir in excitement. He enjoyed the flight, and having beaten the odds. Damn crazy Russians! Jack thought. Technically, Vladimir was Ukrainian and not Russian (just as Jack was Swiss and not German), but he was still eccentric.

Jack calmed down and scanned the surface rushing by below them. The fine 3-D radar displayed the terrain in high resolution color on the view screen. The topography was impressive, one canyon after another bearing witness to millions or billions of years of volcanic activity. They passed a canyon that looked like any water-eroded channel on Earth: it was cut deep into the surrounding rock. Jack instructed Vladimir to follow it in what seemed a downstream direction.

Vladimir turned and said to Jack "What about a little hike?"

"I thought you'd never ask. Let's look for a nice picnic place."

Vladimir reduced speed and looked around for a suitable landing area. The canyon broadened and ended in a plain with broken rocks scattered around.

"That'll do" Jack said and pointed to an area free of large boulders.

Vladimir slowed, engaged the landing thrusters, and set the shuttle lightly onto the surface. He worked carefully through the landing checklist and initiated engine standby mode.

"External sensors read 461°C, pressure is ninety-two atmospheres with a 2 km per hour wind. It's like a Caribbean vacation, no?" He laughed. After a complete shutdown, the astronauts eagerly unbuckled and changed into their cumbersome EVA gear. The heat protection suits were designed to work in an environment that felt like a giant baking oven, plenty hot enough to melt lead. The suits reminded Jack of ancient diving suits. They also had some of the same handling issues. An ancient dry diving suit could send you spiraling up to the surface and kill you by air embolism if you merely pressed the wrong button or panicked: these heat suits could fry you in an instant if you initiated the wrong circulation pattern.

Just as they were ready to leave the shuttle, Jack commented on an unusual energy signal on the scanner. "What is it?" Vladimir asked impatiently while he started to sweat in his EVA gear.

"Wait a minute. There seems to be a radiation source just ahead of us. But I'm not sure. It is very weak."

"I don't see anything special" Vladimir said dubiously as he scanned the surface through a small port window. "But maybe we are lucky and have found some piece of a crashed probe."

"Hmm, doesn't look like that" Jack noted, displeased. "There is some kind of pulsating pattern that extends across several wavelengths. But it is simply too weak to tell for sure. Let's just get out there and take the hand-scanners with us."

"All right" said Vladimir waiting impatiently beside the shuttle door. Finally, he entered the pressure chamber, which, in principle, was similar to those on submarines decades ago, although the shuttles' were much more sophisticated to ensure that none of the outside atmosphere and heat entered the shuttle. That was a major add-on to the costs of these specially outfitted shuttles and had stirred up a lot of controversy years ago. The NASA Administrator at that time didn't think it would ever be necessary for the station inhabitants to visit the surface and didn't want to have to justify the additional cost to the tight-fisted bureaucrats that approved his budgets. Others argued that it would not make much sense to have a Venus Orbital Station without being able to explore the surface of Venus for interesting physical and geological features. In the end, the specially outfitted shuttles were approved, but the crew was reduced to two. A typical bureaucratic solution and compromise— suboptimal in both respects!

When they left the shuttle, they stood in a broad canyon in a bleak volcanic landscape dissected by a few linear features that widened into channels in some areas. The landing site was a flat plain covered with loose rocks and surrounded by hummocky terrain and rounded hills. As they stepped forward, they could feel the hot surface through their boots.

Jack had been looking forward to walking on a hard surface under Earth-like gravity since their last excursion nearly a month ago. It reminded him of the time when he met Amber and they had tried to climb every volcanic ridge between western Arizona and eastern New Mexico. This was different though. The ground felt soft, like walking on a layer of wax, and there weren't many ridges. Venus' temperature would have melted lead to liquid pools if it were present; the heat even softened surface rocks. And with the super-dense atmosphere, even the little 2 km/h wind was so energy-laden that it threatened to push them over. It was like trying to stand up in a slow-moving river: they had to lean heavily into the wind to stay upright.

After surveying the surrounding landscape for a few minutes, Vladimir said, "Okay, what are we going to do now? Where are we going? Do you still have the energy signal on the scanner?"

"No," said Jack. "But the hand scanner isn't as sensitive as the one on board the shuttle. I suggest we walk a bit around in all directions and see whether we pick anything up."

They headed west, along the cliff side. Jack started chipping on the soft rock with a geologist's hand-pick.

"What do you expect to find? Gold?" Vladimir laughed.

Jack did not answer. He just continued for a while, and collected a few rock fragments in his sampling bag. Then they continued to walk together along the edge of the cliff side, always keeping the shuttle in eye-sight. Both Jack and Vladimir stopped intermittently to take pictures. After completing a nearly full circle around the shuttle perimeter, they turned back to the shuttle.

Vladimir suddenly broke the silence. "Any sign of the energy signal?"

Jack shook his head.

After entering the shuttle, they stripped off their heat-protection clothing. It was cumbersome in the cramped shuttle. Vladimir's elbow got twisted in the multilayer heat protection material and Jack had to free him. The air exchange unit was working at full capacity, but the pervading rotten egg smell was simply not getting out of the air supply.

"Doesn't it make you want to gag?" Vladimir complained. Then he took the pilot seat again.

"Well, we still have that weak energy signal on our shuttle scanner. It may simply be a technical malfunction," Jack said.

"We should check it out and fly a circular search pattern close to the surface around our location. If it is caused by something from outside, it should get stronger in one direction."

"Excellent idea!"

They proceeded as Vladimir had suggested and flew larger and larger circles around the landing site.

"I have it!" Jack said. "Go further southwest."

Due to the irregularity of the signal they had enormous difficulties in locating its source, but after about two hours they were in the area where the signal was the strongest.

"It doesn't look like there is anything special out there," Vladimir noted. "There's just a lot of rubble lying around."

Jack nodded. He wasn't sure what he had been expecting, but the endless landscape of broken volcanic rocks didn't appear to be any more exceptional than any other part of the planet's surface he'd seen. Vladimir stopped the shuttle and they both repeated the laborious ritual of putting on their EVA-gear to explore the surrounding area.

"We'll have to leave soon," Jack said as they exited the shuttle. "We are running out of fuel and we are already close to the maximum mission time."

"Sure," Vladimir grumbled into his beard. Since the day cycle on Venus lasted 242 earth days, there was really no problem with daylight. The more pressing issue was possible messages from mission control on Earth while they were reconnoitering on the surface. Mission control could make a big fuss if they were to figure out that Jack and Vladimir were exceeding the maximum mission time without any emergency. They left the ship and searched for anything special that could be related to the cause of the signal.

The landing site, still within the Ishtar Terra plateau, was located at an intersection of two shallow canyons. The flanks on the north sides of the canyons were very steep, with a lot of rubble piled up along them. The larger canyon was about a 100 m wide and the smaller about 50. The canyon's shape reminded Jack of the intermittent streams in warm desert areas on Earth.

"It looks like this canyon has been carved by water a very long time ago," Jack said.

"I thought there is no evidence of liquid water on Venus. Couldn't it be erosion by a lava flow?"

"Before the in-depth exploration of Venus twenty to thirty years ago, the scientific opinion always had been that all of Venus was resurfaced 500 to

700 million years ago. But nowadays most scientists think that there are large areas on Venus that were not covered by any of those lava outpourings, and this area seemed to be one of them." Jack said. "If so, the area in front of us could be evidence of running water from the early Venusian history some four billion years ago.

But Vladimir was more focused on the signal showing on his hand-held scanner than on Jack's explanations. He felt excited, puzzled, and then excited again. Were they on the trail of a big mystery? Or, would this only end in disappointment? He remembered the signal he discovered while working with scientists from SETI a few years ago. At that time they thought that they had discovered a communication signal from an extraterrestrial intelligent civilization and were ready to tell it to the world. But then they found out that is was only the signature of an underground experiment on the Lunar Base. Luckily they had not yet gone public or all their reputation would have been destroyed. But this signal here could definitely not be produced by human interference. Or, could it? Vladimir felt a flutter of uneasiness. The irregular signal had a pulsating pattern, both in energy output and wavelength. It was challenging to pinpoint the exact location. Finally, Vladimir concluded, "It is here, within a few square meters. But I don't see anything except rubble."

They were at the northern cliff side of the larger canyon, about 200 m down gradient of its intersection with the shallow canyon. Vladimir began to move some of the pebbles and cobbles away. Suddenly he exclaimed, "Whatever it is, it must be underneath the surface. Let's get some tools!"

Jack returned to the shuttle and retrieved two shovels while Vladimir excavated small pebbles and cobbles by hand. The shovels started to bend from the unforgiving heat on the Venusian surface that softened the metal. The impacts with the rocks deformed the shovels further. Vladimir sweated profoundly. His heat suit was designed to protect from outside heat, but did not have much allowance for inside heat sources. His visor began fogging up and he kept dripping perspiration on the inside, which severely limited his view. He began to feel light-headed. After more than an hour of digging, they had only excavated a one and one-half meter deep hole from which some bedrock peaked out.

"Look, there is a cavity underneath and it seems to be extending inward and downward. Definitely a cave," said Vladimir and then he took a break.

"Yes," said Jack, who was breathing heavily but continued digging. He was enjoying some real physical exercise, not just the muscle stimulations that they exposed themselves to in the space station. "I bet my back will hurt tomorrow," he said almost triumphantly. "Like I did some real good gardening work. This is fascinating, but I'm afraid that we really have to get back to the station. It doesn't look like we will solve this enigma today."

"Yes, you are right," Vladimir said, but Jack couldn't tell if his comrade was only tired or frustrated. Vladimir's visor was thoroughly foggy and Jack's was starting to fog up, too. Vladimir straightened from leaning on his shovel. "Tomorrow we will have to bring some better tools, maybe even explosives."

"Okay, let's call it a day." Jack wasn't sure if Vladimir was joking about the explosives or just eager to use them.

They assembled their tools and returned to the shuttle. After getting out of their heat protection suits and sitting down, they grinned at each other. The words were written onto their faces. Finally, something exciting!

Vladimir ran through the preflight checklist as the engines warmed up and then boosted the shuttle off the surface with a stomach-dropping thrust. This time the starboard thruster did not give them any problems and they approached the station for docking.

"I think we discovered the signal of one of the old Russian probes that crash-landed on Venus," said Vladimir.

"I don't think so. The pulsating nature of the signal does not make any sense."

"Who knows? Maybe they had a secret experiment or weapon on board."

Jack laughed. "Weapon for what? To shoot some Venusians?"

Now, they both laughed.

When the automatic docking procedure was completed, they checked the mission time.

"Oh, boy! Nine and a half hours," said Jack.

There was a message from mission control on file from five hours ago to be answered in addition to one personal message. "What do you suggest we tell them?"

"Tell them we went for a joy ride," Vladimir joked. "Oh well, you mean whether we should stick to the albedo story or the mysterious energy signal story? I guess we better omit the energy signal story until we have a better idea what it is."

"Agreed. We should phrase it in a way that would allow us to go down again tomorrow. Let's say we checked out a landform due to its unusual albedo and found very interesting associated geological features that need further investigation. That's true in essence, since we found what looks like a lava cave."

"Sure, you are the boss." Then Vladimir went to the shower.

"Don't use up all the water," Jack shouted after him. "You know how slow the recycler is."

## 1.3  The void in the cave

Jack went to the com-screen and established contact with mission control. Mission control seemed to be happy to see him appearing on their screen.

"Where the hell were you?" Fred, one of the controllers, barked at him.

"Oh, sorry. I guess we forgot to tell you about our trip to the surface. And it took a bit longer than expected," Jack reported. Then he told them the story they agreed upon and wished the guys at mission control a good night. Afterwards Jack turned to the personal message. Of course, it was another reminder from his wife, soon to be ex-wife, about the division of their assets for the upcoming formal divorce. When was the anticipated date again? Yes, right, two days before he would return to Earth. What a return that would be. He didn't feel any desire to go back at all. And he couldn't care less about any distribution of assets. Hence, as with the three previous reminders, he ignored and deleted this one.

His thoughts turned toward tomorrow. He got up and searched the station for some better tools, but there wasn't really much around for digging in the space station. They would have to improvise. It puzzled him that he felt so much attraction, so much excitement at the idea of figuring out what this energy signal was. But then it occurred to him that this must be natural given their usual routine on the station which, he had to admit, had become quite boring.

Vladimir returned from the shower. "It's your turn. You look disturbed, what's going on?"

"Ah, it's just another note from my wife about the divorce. I guess I'm wondering what I'm going to do when I'm back on Earth. We leave in less than two weeks!"

"Yes, but before then we will have some fun. Remember, our replacements will arrive in a few days and we will stay with them for one week on the station and supervise them—and they are WOMEN! That should help you take your mind off the divorce."

"I hope so. Oh, we better finish our exploration trip tomorrow, because the day after that we are scheduled to collect data from the probes again to measure microbial population changes. Not that it matters anyhow. There are no major changes really at all in burn-up rates, atmospheric conditions, and so on, since we started taking the measurements nearly nine months ago."

Vladimir seemed to be quite upbeat, Jack noticed, even for Vladimir—and he always seemed to be in a good mood. That was one of his qualities that Jack admired and why he wouldn't have wanted any other companion during the nine month mission. Jack got up and walked silently to the shower.

After the shower he observed himself in the mirror. Truly, he had aged. His hair had become increasingly gray since being on the station. The scar on the left cheek, from a too close encounter with a coral reef during a diving exercise, looked more like an ugly, large wrinkle. The scar used to be attractive to women and he remembered that he liked to make up adventure stories about it. Wrinkles were also developing around his eyes. He looked tired and neglected.

"I am a middle-aged guy, who has nowhere to go!" he thought. "I will turn forty-four next month, but there will be no one to celebrate with me. No wife, no kids, only one sister with whom I lost contact long ago. Well, at least I'm healthy and slim…" he said to himself, "…at least slim as long as I'm on the station. No one can get fat from this food they give us here!" He laughed cynically, more from desperation than anything else.

Jack looked for Vladimir and found him composing a letter to his family, while watching at the same time some kind of comedy show that the Earth Network was broadcasting. Jack envied Vladimir a bit. He seemed to have everything: a happy family that was waiting for him, an enticing career with the Russian Space Agency, and humor and youth, although he looked older than his actual thirty-four years. His beard added some years, but when he was excited and his blue eyes were twinkling, all his youth reappeared. Jack thought that he, himself, had to be on watch so that he would not turn into a grumpy old man. He should probably take Vladimir as an example.

Jack lay down and fell asleep quickly. When he got up in the morning, he felt energized. Vladimir turned towards him and proudly showed him a new digging tool.

"Look what I've built from surplus materials! You know, I wish our station were the size of the Mars Base Station and had all its equipment and supplies. We are so limited here."

"A power shovel. Very impressive. But I don't think we want to use these explosives."

"Never know, they may come in handy."

They had a modest but special breakfast together of oatmeal, dried peaches, and instant coffee, one of the few remaining meal-packages that wasn't the hated beef stew. Then they geared up for the shuttle. The air was filled with excitement. Jack transmitted a short message to mission control indicating the need for further surface exploration, and then boarded the space shuttle. Vladimir waited impatiently in the pilot seat.

"Did you fix the control for the starboard thruster?" Jack asked nervously.

Vladimir nodded. As soon as Jack was in, he released the clamps and the shuttle made its descent. The thrusters ignited without any delay. Vladimir had no trouble finding yesterday's site, and Jack confirmed that the signal was still there. They took out their newly assembled tools, including the power shovel. Jack groaned softly as he stretched his muscles, which were acutely sore from the previous day's exertions, but set about his work with determined and unrelenting vigor. They made quick progress in clearing the entry of the cave: the excavation grew steadily. However, after about two hours, the motor of the power shovel let out a loud pop, with no power remaining, just some sputtering and hissing.

"I guess my heat shield was not good enough," Vladimir said casually.

Still, after another hour they had the entrance of the cave completely cleared of rocks, but some pebbles were still falling from the inside toward the entrance. They took a break and measured the entrance: about 2 m high and 2 m wide. They continued to clear the inside of the cave to see how far it extended into the surrounding basalt. It was amazing how quickly they progressed on the work given the difficult conditions. Enthusiasm and adrenaline!

"It feels like being an old fashioned gold miner," Vladimir joked.

After another two hours the cave was emptied of most rubble. It extended about 5 m into the basalt bedrock.

Jack leaned on his shovel handle and turned to Vladimir with a look of consternation. "We worked so hard and there is nothing in here. It is just a plain old cave."

"I'll get the scanners," Vladimir said.

When he returned, they took the scanners into the cave, and the signal was much stronger. Strangely, it seemed to originate from the empty space at the center of the cave.

"That is the origin," Vladimir proclaimed.

"But there is nothing—just air, plain carbon dioxide," Jack complained, baffled by the unexplained readings.

Vladimir went to the shuttle and tried some scanners for temperature, wind, magnetic and electromagnetic readings. But none of them indicated anything unusual. They went back to the shuttle, and moved it with its on-board scanner into one line with the cave, which confirmed the energy source being in the cave and most likely in the center of it.

"Maybe we're going after it the wrong way," Jack said. "Maybe we should look to see if we can identify a pattern in the energy pulsations."

"Yeah, right," Vladimir said. "The intelligent Venusians are trying to communicate with us, from out of thick air." It was unusual for Vladimir to make such a good pun in English, and he laughed at himself.

Jack started analyzing the irregular pattern of the energy source, while Vladimir took out every additional scanning instrument that he could find in the shuttle to analyze the center of the cave. After nearly an hour of silently doing their work, they met back in the shuttle to share results. Vladimir spoke first. "What did you find?"

"Nothing really. I looked at the signal for nearly an hour and tried to figure out a pattern with various algorithms, but I couldn't see anything. It appears to be mostly random, but contains some repeating patterns. That doesn't make any sense. There must be a reason for the signal to be there."

"Can it be anything artificial? Perhaps from one of the previous probes that crash-landed?" Vladimir suggested.

"Not really. What would it be? The cave is a natural lava cave. There is no artificial debris around whatsoever. And the signal comes from the center of the cave. It simply doesn't make any sense."

Then Jack, looking hopefully, asked Vladimir "What did you find?"

"Nothing much either. All environmental parameters that I measured at the signal source-point appear to be the same as in any other location of the cave. The cave environment is a bit different from the surrounding environment, but not more than expected for a cave. One thing though was puzzling. At the center of the cave, where we think the signal is originating, the air density is less. I re-measured it several times, but there seems to be a definite drop."

Jack smiled. "There is nothing more intriguing than a good mystery. I just wish we could have a bit more to go by. But what I can do is to take apart the environmental sensor in the shuttle, which is much more sensitive than your hand-held instrument, and see whether there really is a drop in air density."

"Yes, let's do that. I'll help you." It took them about half an hour to take the sensor out. Then they suited up again, connected the sensor to the outboard power supply, and carried it to the center of the cave.

"No air density differences," Jack commented, now somewhat confused. Had he made some stupid error earlier?

Vladimir took a step back to get some perspective on the situation. "We are off from the center of the cave. We have to move a few centimeters toward this wall."

When they finally adjusted the position, the air density measurements were one tenth to one hundredth of normal.

"Hmmm," Jack mumbled. "There is definitely something. The sensor takes its reading from a cubic millimeter of air, and depending on how we adjust it we do get a ten to one hundred fold drop in atmospheric density. It seems to be real! I wonder if we could take the reading from a smaller volume still, and see whether the effect would become even stronger."

"Can you do it?"

"Well, I think so," Jack said confidently. "But not here. I would have to work on that when we are back in the station."

"It's late anyway," Vladimir said. The sparkling in his eyes was gone and he looked tired. "Maybe we should call it off for today. We can't really do much more here without getting better equipment."

"All right," Jack agreed, but was obviously disappointed that they couldn't find out more during their presence on the surface. He knew his friend was right. He just hated having to leave this unsolved puzzle behind. Their time on the station was running out fast.

They collected the equipment they had scattered around the landing site, went into the shuttle, and returned to the station.

When they sat together in the evening, Vladimir asked, "What is the plan for tomorrow?"

"Well, tomorrow we have to monitor the microbe probes again. But I can feed the data I collected on the pattern of the signal into the computer, and we will see what it spits out. Meanwhile, in my spare time, I will work on a more sensitive sensor for the source of the energy signature."

## 1.4 The Discovery

It was only two more days until their replacements from Earth would arrive. They had looked forward to this change of routine for weeks if not months. But now that they had discovered the intriguing phenomenon, time seemed to run out far too quickly. Jack wished that the replacements would come at least a week later so he and Vladimir would have sufficient time to thoroughly investigate the puzzling emanations from the cave. Two days was definitely not enough time. Even worse: today, microbe monitoring was on the schedule, after which Jack would have to compile all of the monitoring data. They would have even less time to investigate the phenomenon. In the morning they took the shuttle, checked the measurement locations, and took readings of the number of microbes counted, the routine atmospheric conditions, and other ambient chemical and physical conditions. When they arrived back at the Venus Orbital Station, Jack started to put all the data together from all their microbe sampling events going back to their first sampling nearly nine months ago, including today's data. This was the last measuring event they were responsible for. The next one, in three days, would be the responsibility of the replacement crew. Jack examined the data for each measurement probe, going back to the nearly twelve years of records since the Venus Orbital Station was functioning and recording data. His final report on their scientific investigations was due before returning to Earth, and he always liked to get things out of the way quickly. When going over the microbial counts, he noticed a definite cyclical behavior. He downloaded the data taken by previous station inhabitants. Yes, when he put all the data together, there were patterns, small cycles overlying medium cycles overlying larger cycles. The small cycles were in the range of about two to three weeks, the medium cycles in the range of thirty-five weeks, and the large cycles in the range of about five years.

Jack got more and more into the calculations and attempted to make some sense of the clues in front of him.

Vladimir popped in to Jack's working area. "You seem to be busy."

Jack took a deep breath and pushed himself away from his computer. "Yes, the data actually seem to reveal a pattern."

"Let me take a look!"

"Later," Jack dismissed his friend. "I want to check them one more time. Why don't you take a look at the pattern of the energy signature that we picked up on the surface? We are still on to go down tomorrow. Are you ready?"

"Sure," Vladimir responded, and went to the computer where they had stored all their surface readings from the cave.

Jack analyzed and re-analyzed the microbe data—microbial populations versus time. There was definitely a pattern, but the large pattern was difficult to nail down because the data available was only twelve years total, so the five year cycle could only be a rough estimate, at best. The small cycle of two to three weeks was somewhat problematic as well, since they took readings only every three to four days, and in the past the frequency had been even less. And then there were data gaps because of technical problems and such. Nevertheless, the cycle was somewhere in between two and three weeks. This left the medium cycle.

Jack felt the anxiety rising within him. He hated to leave things unfinished. He simply was incapable of letting things go or of quitting. This was probably the reason why NASA valued him—why he was the head scientist on this mission. On the other hand, in his personal life this attitude definitely contributed to the demise of his marriage. Jack grew more impatient by the minute. He ran another statistical analysis. The computer displayed a cycling pattern at a time interval between 239.8 and 243.1 Earth days at a 99 % confidence level.

"This can't be a coincidence," Jack spoke aloud to himself, excitedly. "The day-night cycle on the Venusian surface is 242 Earth days."

He re-ran the analysis and came up with the same result. Then he started plotting microbial populations against time. Even with the naked eye he could see that there was a pattern and the 242-day cycling came out very nicely once he knew where to look. But he was still doubtful. Why should there be a correlation between microbial densities in the lower atmosphere and the day-night cycle on the Venusian surface? He sat back for a minute and poured some coffee powder and hot water into a cup. He sipped and let his thoughts drift.

Finally, a possible useful finding from all this boring monitoring, he thought. Then he looked at the graphic displays again. Somehow, they looked strangely familiar. A thought shot through his head. He ran into Vladimir's compartment. Vladimir was lying on his bed and looking at the energy signals they had recorded from the cave.

"Let me look at those!" Jack said, and grabbed the display.

Vladimir looked puzzled, but moved to the side. Jack looked over the signals, changed displays and looked over them again. Vladimir sat patiently until he had to interject something: "Can you finally tell me what this is about?"

"One moment. One moment," Jack said quickly as he whipped through some hasty calculations.

Then he turned to Vladimir. "Tell me that I'm crazy. There is a cycling pattern in the microbial densities from our microbe monitoring. The energy signatures that we took in the cave seem to mimic the same pattern, but with a much shorter wavelength."

Vladimir looked concerned.

"Why don't you look at the data and see whether you can find the same pattern or whether I'm only totally overworked and starting to hallucinate?"

"All right, I will. But just for the record, I know that you are overworked so I slipped something into your coffee today to make you hallucinate."

Jack looked at his friend and Vladimir's face was serious, but then he broke into a grin. "I'm just kidding. But you *are* working too hard. I will take a look at it and see if I can see the same thing."

Jack went to his bed and started thinking about the implications. His mind was buzzing. He waited 10, 15 min—no sign of Vladimir yet. He started walking up and down in his compartment, but no, he should simply wait for Vladimir to come to him. He wanted to make sure that whatever he found could be found by other scientists as well. The data would not lie.

After another excruciating 20 min Vladimir stepped in with a large grin on his face. "Well, either this is the Nobel Prize or the mad house for us."

"Did you find a pattern?"

"More than that, I overlaid the pattern from the microbial density with time to the repeating pattern from the energy signatures with time. Come, look!"

Jack followed Vladimir to the computer terminal.

"Look, Jack, there is a cycle that fits both patterns. The only difference is that the energy signatures are on different time scales. But look—it is the same pattern!"

"Let's have the computer analyze it," Jack said excitedly. He plugged in the commands hastily and then sat back as the computer worked through the analysis. The passing seconds seemed like an eternity.

"There it is!" Vladimir exclaimed.

A 99.78 % match at a 95 % confidence level, the screen displayed. Vladimir grinned and Jack sighed. They sat back, lost in their thoughts.

Vladimir broke the silence first. "What do we do? Send our analysis to mission control?"

Jack shook his head. "They will think we are nuts." Finding the pattern was the easy part. To convince mission control that this is real will be the difficult part. And then to figure out what it meant!

"But we have to tell them eventually. Even if they think we are nuts."

"Yes," Jack agreed, "...but let's get more information first."

"We don't have much time tomorrow. We have to prepare the station for the replacement crew."

"Yeah, we probably should just spend tomorrow preparing for our replacements and re-analyze the data. We could get them involved and the following day all take a trip to the surface. If we can convince them, maybe we have a shot with mission control."

"What if they claim this extraordinary finding for themselves? Can we trust them?" Vladimir looked concerned.

"There is really no danger. I'm formally in command of the station until we leave. Let's get down to the surface the day after their arrival. If we get confirmation and can convince them, we send off a report to mission control right away and there will be no doubt who made the discovery. If our theory falls apart, well, we saved ourselves from embarrassment. As I see it, a win-win situation for all concerned-especially you and me!"

"It would also give me some time to develop a more sensitive scanner."

"Good. But let's find out as much as we can about our replacements!"

They spent the next two hours getting to know everything they could about the astronauts slated to join them on the station. Vladimir cruised through the personnel webpage of NASA and the European Space Agency. Vladimir was familiar with Ludmilla. He met her previously at mission planning events, but had only engaged in some small talk with her. The posted bio on the ESA website revealed that Janine Ludmilla was born in Bulgaria forty years ago and received several science awards and honors even while still in high school. He recalled that she told him about her studies at the Russian Academy of Sciences and her year at MIT. Vladimir wondered just how trustworthy she would be, it they needed to keep this discovery secret. The 3-D hologram that plopped up on their computer screen showed a determined woman with a narrow mouth, blue-greenish eyes, and blond hair. Vladimir tried to recall the impressions he had when meeting her. Yes, Ludmilla seemed to know what she wanted and was apparently up to any challenge. Perhaps, she could even help them to solve the mystery. Janine was listed as a designated astronaut of the Russian and European Space Program and clearly one of the rising stars of Europe's new planetary exploration program.

Jack found Carmen Mejilla on NASA's personnel webpage. Her photograph took his breath away. She was stunningly beautiful. He swallowed: too long without a woman, he thought. He ripped his eyes away from the photo, and scanned the page. Carmen was Venezuelan, age 34. She was born and grew up in Tucson, Arizona. Jack couldn't find any awards of excellence in her record, but Carmen seemed to have a talent, to always pop up in the right place at the right time. Her studies focused on microbiology and had been tutored by some of the best scientists in the country. She apparently didn't attend many scientific conferences, which explained why he or Vladimir had never met her. Carmen had a ten year old daughter, which probably slowed down her career. Or maybe she was not as career-driven as he was, Jack thought, as he rubbed his forehead over and over again. He entered the name of her daughter into the search engine. Ah yes, her daughter lived with her grandmother near Pasadena, while Carmen was on the Venus mission. Jack stopped for a moment. That was getting too personal! He reloaded Carmen's webpage and his eyes swung back to her photo. He probably could just have stared at it for hours. In the holographic picture Carmen had a wide smile, brown eyes, and shoulder-long brown hair. Somehow she looked familiar, yet he could not remember ever having met her. And he would certainly remember that! Then it occurred to him. Yes, yes! Wasn't she the woman in his dream, the one screaming for him while he was struggling to get out of the black hole and explosions pummeled Earth?

"Quite an attractive replacement," Vladimir commented as he walked over to Jack's computer screen and saw Carmen's image.

Jack, torn from his thoughts, quickly regained his composure. "They're certainly better looking than us," he agreed.

"And better groomed, too!"

Both men laughed.

Then they went to bed, each of them thinking about the events of the last few days and the coming arrival of the two women.

## 1.5 Carmen

The day started with a haze that seemed to be denser than usual around the orbital station. Gases from the surface produced increasingly thicker clouds in the lower atmosphere. But Jack and Vladimir did not pay much attention to the changing atmospheric conditions. They analyzed and re-analyzed all their results, but came up with the same conclusions. With help from Jack, Vladimir improved the sensitivity of their scanner by a factor of four.

In the evening they finally decided to clean up the station. After all, their replacements were scheduled to arrive tomorrow.

"It reminds me of cleaning up my room when I was a little boy, because my mother wouldn't allow a messy room," Vladimir said, but pitched in to do his part.

Vladimir was in the CCS when they received a brief live transmission from Janine Ludmilla that they would arrive according to plan. The arrival time was also confirmed by Mission Control on Earth. They received no other news. Jack was surprised that there was no personal message from his soon-to-be ex-wife. For sure, she must be steaming with anger by now, because of me ignoring her, he thought.

Vladimir spent at least an hour communicating with his family. After all, only one more week on the station to train the replacement crew and then twelve days for the flight from Venus to Earth, and he would be home. He was clearly excited about returning. Well, for Jack, there was not really a home anymore. His home was space and planetary exploration.

"Discovery IV ready for docking," a soft voice sounded through the communications link. Vladimir was first in responding, "You are a bit early. Sure, we are ready."

The docking to the engineering module went smoothly. When the door opened, the two new astronauts climbed out of their transport vehicle. First, Janine, then Carmen. Jack realized how neglected Vladimir looked when noticing how attractive Carmen appeared with her brown tinted skin and shoulder-length brown hair. Then, he realized that he probably did not look much better than Vladimir after nine months on the station. He quickly pushed such thoughts aside and settled back to his duties.

"Welcome to the Venus Orbital Station. We are looking forward to some company," Jack said.

"Thank you," Janine replied. "Greetings from Earth."

Then they walked to the control room. Vladimir and Jack felt a little bit awkward. They didn't know what to say. It seemed that they had lost some of their social skills and familiarity with human beings during the last nine months.

"Well, this is your new home," Vladimir broke the silence. "We hope you will enjoy your stay."

"I'm sure we will," Carmen responded. "We brought something special for you. Some real chocolate. Swiss chocolate."

A wide grin spread over Jack's face and he whooped in excitement. Sweets were his passion and chocolate from his old home country was his favorite. They sat down with station coffee and chocolate, and talked about their nine month's of experience on the station. At the end Jack and Vladimir, in a mixture of excitement and nervousness, hesitantly shared their findings about the energy pulse on the surface and its correlation to the observed microbial populations.

"That's fascinating!" Carmen said.

"We should look through all the data and analyze them again," said Janine, somewhat critically.

"Yes, you definitely should!" agreed Jack. "And after you have done that we should go to the surface so you can check out this strange cave for yourself. We will take the improved scanner with us."

"Sounds good," Janine replied, with some curiosity now in her voice.

*That was easy*, thought Jack, but after all, he was still station commander until they left the station, at which point he would pass that command formally over to Janine.

The next day the space station was humming with activity, with all four people moving around getting their gear together. There was a line at the shower that morning. When Jack arrived, Vladimir and Janine were already waiting for Carmen to finish. Jack grabbed the shuttle checklist while waiting. But his focus was elsewhere. The scent of the feminine soap and fresh body smell from the women made him remember the exotic honeymoon he had with Amber so long ago. They went to one of those luxurious places on Tahiti and went diving and surfing all day. It was a dream but it stirred him and made him uncomfortable in a nice sort of way. His thoughts were interrupted when Carmen emerged from the shower only skimpily covered with a bath towel. Jack couldn't help staring. Janine entered the shower next and Carmen squeezed by in the narrow corridor. He didn't know what triggered it, but pages from the shuttle checklist that he held in his hands suddenly slipped out of his fingers right at the moment when Carmen was passing. He grabbed for them in a reflex, but lost balance and bumped into her. Part of the towel came off for a moment and he could feel his skin on hers. It sent shivers up his arm and other places.

Vladimir, who noticed the incident, said jokingly, "Wow, Jack, you don't lose any time, do you?"

Jack felt how his head heated up and thought that he surely must be as red as a tomato. You are such a clumsy idiot, he told himself, and then, even worse, you get red like a little school boy. Jack just mumbled "sorry", and didn't say anything else. What could he say?

A quick smile graced Carmen's lips and then she vanished into her quarters.

Jack pushed the unpleasant, embarrassing thoughts away. He sorted the checklist and settled back on the tasks. The women had to be instructed in shuttle operation to fly down to the Venusian surface. He once again became the consummate professional that he was, the commander of their mission and natural born talent in scientific reasoning and leadership.

Mission briefing. Jack sat in the CCS module with Vladimir, Carmen, and Janine and went through their planned mission to the Venusian surface step by step. "You, Carmen and Janine, will have to take the number two shuttle that is anchored to the station: it's usually reserved for emergencies. I believe this is a special situation and justifies its use."

"Why don't you use the actual emergency shuttle?" Janine asked.

"We, ahem, had a technical problem a few days ago with the other shuttle and I'm afraid only Vladimir has a sensitive enough touch to get the controls un-jammed if that were to happen again.

Carmen looked somewhat nervous, Jack could tell. She probably did not expect to visit the surface of hell on her first full day on the station. But she wasn't saying anything, she was a professional.

"Vladimir, please go with Janine and Carmen through the shuttle instructions one more time, especially the drop mechanism. Meanwhile I will send the flight plan to Earth headquarters. We leave at 9:30 sharp. Any questions?"

The women shook their heads.

Soon afterwards the astronauts started stowing their gear into the shuttle. The shuttle now appeared even smaller, but there was no other way down. Jack could just wonder how anxious the women would feel. He felt responsible and a bit uneasy about rushing them right away down to the surface before they even had settled into the station. The thing, or, better, LACK of thing, down there on the surface had a near-magical attraction to him. They had to solve the mystery and going back down was the only way to do it!

Janine and Carmen boarded their shuttle and the men heard their "ready for descent" through the com-link. Then both pilots executed the drop mechanism. Vladimir was able to ignite the thruster flawlessly this time, but Jack was only relieved when he received the notification from Janine that their thrusters were working properly as well. Cross-winds picked up near the surface and shook the two shuttles like a misaligned roller coaster on its tracks. The women, despite being less experienced with this type of shuttle, kept up with Vladimir's piloting, and they all finally reached the landing location near the cave.

They geared up. Then they climbed out of the shuttle and walked toward the cave. Jack caught a glimpse of the tenseness in Janine's face even though it was half-covered in the EVA-suit. They are definitely shaken, Jack thought, but they are holding up remarkably well. When they reached the cave, Carmen broke the silence first.

"I don't see anything special, just a lava cave," she said doubtfully.

"You won't see anything special," Vladimir responded. "It's all in the readings." He switched on his sensor with the improved sensitivity. The new sensor worked great. Whenever Vladimir pointed the sensor into the center of the cave, he registered the energy emissions.

"Very, very low air density associated with the source," he emphasized.

"I can register the energy pulses on my sensor, too," Janine said, who held the standard hand sensor from the shuttle. "I think we should mark it."

"Excellent idea," Jack said. "But how do we mark something in the middle of the air?"

"Maybe we can use two focused flashlights or lasers and have their beams intersect at the source," Vladimir suggested.

"Wouldn't that affect the energy source?" commented Carmen.

Vladimir and Jack looked at each other. Then, Jack moved toward the source and moved the beam of the flashlight through it. "Did you measure anything different?" he yelled to Vladimir.

"No, it didn't affect the energy pulse at all. And it didn't seem to affect the low density air detection either. Do it one more time!"

Jack repeated moving the beam of the flashlight through the source. Then he moved the flashlight itself through the source.

"Anything now?"

"No, this is very odd. The low density effect should be gone. After all, the flashlight has a higher density than the air. But I still get the same reading."

"Maybe your sensor is broken," Janine commented, who was slightly distressed about their experiment.

"No, no, it works fine. See!" Vladimir pointed his sensor into different directions.

"Okay, why don't we try to measure how big the source is?" Janine said.

"I was afraid you would suggest that."

Vladimir started to outline the size. The others watched him and stared into thin air. Then Carmen and Janine began to study the energy signatures that they received through their sensor. Janine, at one point, went back to the shuttle and used the on-board sensor. She picked up the signal with that one as well.

"Come on, Vladimir, let us know what your miracle sensor tells you," Jack said impatiently after about half an hour.

"Well, you know, it is difficult. Of course, I have to run it through the computer, but as it looks right now anything from the size of one cubic centimeter and a density of one one-hundredth of the atmosphere, about 0.8 Earth atmospheres, to a size of about a cubic micron or so and very low density, perhaps even absolute vacuum, seems possible.

"I think it must be fairly small," Carmen commented, "otherwise there would definitely have to be some detectable interactions with the flashlight."

"She has a point," Vladimir said, and Jack and Janine agreed.

Jack took a small pipette out of his tool kit and picked up a small pebble from the ground. Then he moved it through the center of the energy source.

"We will take this pebble with us and examine it on the station under the microscope. Hopefully, we can find some kind of effect that we can trace. Any other ideas?"

The women shook their heads.

"Oh, we should do the same with a few more pebbles," Vladimir hurried to mention. Each of them picked up a few pebbles and moved them through the source.

"So, I guess everyone has their one souvenir," Vladimir grinned, and added "I still don't see any change in energy pulse or change in air thickness."

The crew returned to the shuttles and flew back to the station. Upon return, the two microscopes on board were occupied continuously.

Vladimir got frustrated first. "I don't see anything at all!"

Carmen took over Vladimir's microscope and used her pebble. First she thought that she saw something, but then realized that it was only cooling cracks resulting from the temperature difference between the Venusian surface and the room temperature on the station. She continued looking. Meanwhile Janine had no success and finally let Jack use her microscope, and mumbled, "I'm not a petrologist. I don't even know what I'm looking for."

Jack began to get frustrated as well. "There certainly isn't anything big."

Then, suddenly Carmen said in her characteristic soft voice, "I think I have something. There is a space just below a micrometer in size in which the mineral structure looks rearranged. Like its might have been taken out and then emplaced back into the pebble. The disturbed materials go through the entire rock and have absolutely sharp edges. It's certainly not anything that seems to be natural."

Jack looked for the same feature under his microscope.

"Now I see it too," Jack said triumphantly.

Then, he let Vladimir and Janine look at it. "That definitely looks odd," Janine commented.

Vladimir jumped up. "How can that be? It is like the energy source or whatever it is, melted a hole through the whole pebble and then inserted the atoms back into their original space, but offset and in perfect arrangement! But in a different arrangement!"

"It's not melting," Carmen said. "That would look different. It is more like the atoms were simply removed. One other thing is odd, though. Did you notice that there are no impurities in the crystal structure in the area that was replaced?"

"Yes, that is true. Any mineral structure contains impurities all over, but the replacement area does not. Is it possible that the replacement path was coincidently pure before we moved the pebble through the source?" Janine said.

"I think that's extremely unlikely. But, of course, we can check on that possibility by looking at the other pebbles."

They did as Carmen suggested and found the same results for the other pebbles.

Later that evening when they sat down, Janine insisted that they had to report their findings to mission control, but Jack was hesitant. "They will not take us seriously at mission control!" he said. But after an hour of discussion he finally agreed. "All right, I'll put a report together tomorrow morning, after I get some sleep." Then the group trickled away and went to sleep. Jack realized how much more lively their station appeared with the two women on board. He definitely enjoyed their company, especially Carmen, who appeared to bring some sunshine into his life. He would feel comfortable to have that going on for a while, but then Vladimir and he had to depart shortly. It was better not to get too attached to the current situation, he thought. His extended isolation on the station had probably altered his perceptions anyway.

## 1.6 Death is final

Early the next morning Jack and Carmen met in the control room.

"Would you like a cup of coffee? It's not great, as you know, but it's the best we have around here and you get used to it," said Jack.

"Sure," Carmen replied. "How does it feel to be on this space outpost for nine months?" she asked.

"Oh, well. Somewhat dull at times. Sometimes lonely, being so far away from anything. But then the last couple of days have been real exciting!" Jack said wistfully. "Besides, I like space. Up here, there seem to be so few problems; just you and the universe. It's somewhat romantic in a way."

"But don't you miss your family? Your friends, Earth? I miss my daughter already. But I guess you should not have a family if you are an astronaut. That's what my ex-husband told me when he left." Carmen swallowed.

After a small pause, Jack responded. "I don't know what the answer is. I guess at some point one has to decide where your life is, in space among the stars or on Earth."

Both stared out of the window: the clouds had lifted and dispersed a bit and they had a stunning view of the Venusian surface.

"But here," Jack said with a smile, "here you have a whole planet at your personal disposal."

Carmen smiled back and Jack regretted that he could not spend more time with her. Jack felt how his head heated up again. Perhaps there would be some possibility of developing a closer relationship with her? His thoughts were interrupted when Janine entered the control room. "A great view!" she said.

Again, Jack felt like a school kid, a school boy that was caught doing something he shouldn't. "Yes, you won't have this kind of view every day. Enjoy it." Jack replied. "Well, I better work on the report to headquarters about our Void and its energy signatures."

Carmen and Janine returned to unloading the rest of the equipment and supplies from Discovery IV into the Venus Orbital Station. Vladimir helped them out and noted, on the food manifest, the complete lack of beef stew. Maybe the mission-control people had learned something from his complaints? At least the women would be spared of that, he thought. Meanwhile, Jack was putting the report together describing in detail their observations, both on the surface and in the space station. He also included the energy signatures of the Void and its correlation to the microbial life cycles that they had discovered. After he was done, he gave the report to the others to read. They all agreed that it looked pretty good and was objective. Jack sent off the transmission and suggested that any further questions could be addressed during their upcoming scheduled teleconference in six hours' time. Mission control acknowledged receipt, as they did for any routine message.

"Did you hear anything from mission control?" Carmen asked Jack when they met in the control room again.

"No. They probably don't know what to make of it. Or, they are simply waiting for our telecon."

"What the heck?" said Vladimir as he entered the control room. "We know what we saw and measured, and it will make a great paper in some really fine journal. And we're going to write it on our ride back to Earth."

They were all assembled in the control room for the telecon. Jack established contact with mission control.

"How are you guys doing?" Fred greeted them. "Everything fine?"

"Yes, great. All are in good health and spirits" Jack responded.

Fred continued with some small talk, which started to annoy them. Jack finally interrupted Fred. "What do you think about the report that we sent you a few hours ago?"

"Very interesting," Fred responded. "Wait a second. Mike Hang wants to talk with you about it."

"Wow!" Vladimir thought loudly. "The NASA administrator and Deputy Chief for International Space Exploration. Those guys are taking our discovery seriously!"

A white-haired energetic man appeared on the screen. Mike Hang had been appointed to the position only six months ago, but had already made his mark within the structure of the Agency. In most circles he was considered a rough-neck, but also a straight shooter, who said what he thought. Many scientists, engineers, and astronauts admired him for that, especially since the former few NASA administrators had come from the business community or–worst of all—had been merely politicians. On the other hand, many scientists were worried about his religious beliefs, which he acknowledged with unusual openness. He was not very popular within some political circles, but it seemed he had the backing of the President and other highly influential people.

"Quite a find you made there, Dr. Kenton," he addressed Jack. "What do you think it means?"

"I honestly don't know," Jack responded. "We are still just describing the phenomenon."

"Dr. Ludmilla, do you agree with the finds of Dr. Kenton and Vladimir Kulik?"

"Yes, Sir," she agreed.

"Can't these signals be caused by malfunctioning instruments?"

"No, we double-checked all systems. We have multiple readings from different types of instruments. This is real, there is no doubt," Janine responded.

Vladimir and Carmen nodded.

"Well, alright then, Dr. Kenton. I'm looking forward to seeing you upon your return from Venus. And I might add, I believe that I have a new assignment for you," and then he hesitated a bit and added, "If you are ready."

Mike Hang disappeared from the screen and Fred re-appeared: he started chit-chatting again. Jack got tired of it, but the others were involved, too, and he thought that a bit of light conversation wouldn't hurt. Finally, even Fred seemed to run out of material to talk about. Somewhat suddenly he said "Jack, I need to talk with you in private."

"We are all family here. Whatever you have to say you can say to all of us," Jack responded.

Fred insisted: "I really do think that this should be only for you."

The others left the control room without a word. "Okay, I'm alone," Jack announced.

Fred, in a somber voice, said, "I know your wife and you were not on the best of terms. But there has been an accident." Jack felt his stomach twist into a knot even though he tried to stay calm. "She was in a car accident and died in the hospital this morning. I'm very sorry, but I thought that you should know it right away."

Jack felt the blood filling his head and then draining out again. With his face now pale, he thanked Fred, stared for a moment at the screen, and then turned off the telecom system. He continued staring for perhaps another ten or fifteen minutes, while various memories rushed by. These were mostly good memories intermixed with sad feelings about his lack of understanding why things went so sour. He always thought that he had tried everything, but now he doubted himself. Had he really tried everything? Should he have given up when he did and signed on to this mission? Although he thought he had completely disconnected long ago from both her and the relationship, he was troubled by how much this message shocked and unsettled him on a very deep and basic level. This was the news of no return. Whatever he had failed to do in the past, there was nothing he could do anymore—no remedy, ever.

Perhaps there had still been, someplace in his mind, the hope that they could get back together and things could be great again, like early in their relationship, but this glimmer of hope was extinguished now. Death was final. Eventually he rose from his seat, feeling aged by a decade or more, and walked toward his personal quarters.

On the way, he saw Vladimir and Carmen in the corridor. Vladimir approached him concerned. "Is there a problem between you and mission control?"

"No," Jack replied flatly. "It's personal." He paused for a moment. "My wife Amber was killed in a car accident this morning." He was trying to be strong but his words sounded flat and uncaring. How the hell was one supposed to respond to the death of their spouse? Jack didn't know whether he should be strong or just let out his fear, hurt and confusion.

"Oh, God. I'm really sorry," Vladimir said.

Carmen and Vladimir hugged him spontaneously. Vladimir mumbled something that Jack couldn't understand, but he tried to focus on maintaining his composure. All he wanted to do was escape the moment, so he quickly excused himself and moved on to his sleeping area to be left alone with his thoughts.

## 1.7  Aftermath

Jack didn't feel like interacting with the other crew members for the next few days, not for conversation, or anything else. Nevertheless, the days passed rather fast with everyone preparing for the upcoming changes. Discovery IV had to be made ready for departure and Carmen and Janine had to be familiar with all aspects of handling the orbital station. No one brought up Jack's loss. The focus was squarely on their duties. Finally, the hour of departure came.

"You have everything?" Jack asked Vladimir.

"Yes!" he responded. "And I have downloaded all the data to put together a fine paper."

"I wish you a good stay on the station," Jack said to Janine and Carmen.

"Thanks!" said Janine and shook their hands.

Carmen gave each of them a good-bye hug. The hug made Jack feel slightly awkward—even this low level of intimacy and familiarity made him feel uneasy. His thoughts drifted briefly to his wife. Was he feeling guilty? He wanted to say something to Carmen and fumbled awkwardly to think of something appropriate and professional. "I would like to stay in touch with you," Jack finally said. Darn, that was neither appropriate nor professional.

"So would I!" she responded softly.

Then the doors closed and Discovery IV undocked from the Venus Orbital Station.

Discovery IV was a bit larger than the station shuttle, but Jack still felt cramped, especially with all the supplies and miscellaneous materials that they were carrying back to Earth. He didn't mind that much bone-bending accelerations or stomach-churning turns, but hated being confined to small spaces for long periods. The small blue glittering marble in the front window that represented Earth grew in size every day. Yet, there was no hurry for Jack to get back. He didn't know what to feel or think. Amber had been a constant source of anger and frustration for the last two years, but at least in some weird way she gave him a sense of stability, some kind of connection with Earth and all the other humans on the home planet. Something he could hang on to. Now he would soon be back on Earth and would feel much more empty than he felt in space. Nothing to attach or cling to.

Vladimir and Jack spent most of their time writing the paper. Jack wasn't too excited about the scientific article either, but at least it kept him focused on work rather than thinking constantly about his wife's accident. Vladimir thought a lot about the coming reunion with his family, and pushed Jack to have the paper finished upon their return. He wanted to take a long break from everything that had to do with science as soon as he stepped on Earth's soil. Otherwise, Jack and Vladimir did not talk very much. Their emotional states were just too different and each of them knew this and understood it.

# 2 Earth

## 2.1 The Complex

"Welcome back," Mike Hang greeted Jack in his office. The NASA administrator appeared to be genuinely happy to meet Jack in person. "Must have been a long trip," he continued. "I expect you'll be happy to get rested up back here on Earth."

"Actually, I would be grateful if I could start a new assignment as quickly as possible."

"Well, you are "The Man" right now. Seems you've made what may turn out to be the greatest discovery of the century. And we would like you to continue investigating this new phenomenon, if you'd like to do so. But before…"

Jack interrupted him. "You mean, you want me to return to the Venus Orbital Station?"

"No," Mike said with a smile and sipped on his tea. "Please sit down." Then he continued. "The discovery has not been publicly released yet since it has not been confirmed, but these kinds of energy signatures that you found, well, we've detected them in other places as well."

"How can you have discovered them so quickly?"

"That was fairly easy. We just screened areas, looking for the frequencies that you provided us for the Venusian Void, and looked for very similar pulsating patterns. We found them quickly. Especially the one on Earth. It is extremely strong."

"You said "they"?" Jack's eyes widened in astonishment. "Plural? So there are more?"

"Well, here on Earth and also on Mars, so far. The one on Earth is much stronger and the Void much larger than the one you discovered on Venus. The Martian version is slightly larger and stronger than the one on Venus. And then we detected a matching signal from Titan—apparently Titan's Void is larger than the one on either Venus or Mars, but still much smaller than ours here on Earth. Also, we detected another one that is totally puzzling to us and appears to be coming from the Kuiper belt."

"From the Kuiper belt?"

Mike shrugged. "It appears so. At least that's where it was. It's apparently moving toward the center of the Solar System and it's currently located between Neptune and Uranus. Powered by exactly what, we haven't a clue. But our resolution is so poor that this could be all about nothing. However, if it is confirmed, the signal must be strong and the pulsating pattern must be quite a bit different from that of the other Voids. But for the moment let's focus on what we are reasonably sure of. The one on Earth is huge compared to the one on Venus. Nearly ten centimeters in diameter, and with a signal about 100,000 times the strength of the Venus signal. One of our scientists, Dr. Zadak Szodan, speculated that a Void's size and signal strength might be related to the biomass on any Void's planetary body."

He paused and then continued. "We are currently focusing our efforts on Earth, examining the energy signature with everything we've got. On Earth the Void is located about 15 m below the surface. In Jerusalem, on the Temple Mount underneath the El Kas Fountain, a religious place for three major religions, which makes any access incredibly difficult. Nearly impossible, as a matter of fact. Can you believe this? Of all places! Some of us want to study the Void using deep probing and imaging techniques. Others propose excavation to reach the bedrock beneath the fountain. But any invasive technique is out of the discussion right now. Otherwise we'd quickly have riots on our hands. As you can imagine, that's just about the worst place imaginable from a political standpoint. Accessibility is extremely limited and any field studies or experimentation is bound to be under heavy scrutiny. The implications of the location are more than enough reason to proceed with the utmost caution.

Even though we didn't announce anything publicly, from what leaked out we already have a faction that doesn't want to disturb anything. They are afraid that any tampering with the Void may affect life on Earth, since that thing or whatever it is, seems to be related to life or biomass."

Jack was intently listening, but gauging his tone and posture carefully. Mike Hang was known to be, by some estimates, overly religious. Jack wondered if Mike was having any personal internal conflicts between his faith and his duty to NASA and commitment to science. If he did, he gave no indication. Jack let the thought go and focused on the incredible news.

"If it is so big, why wasn't it discovered before?" Jack asked.

Mike shrugged his shoulders. "Well, I guess, nobody looked in that frequency range and that location, or maybe somebody did notice it before and then casually disregarded it as an anomaly. That's why it went unnoticed until now. In some way it is quite funny. We had it right in front of our nose, but never noticed it. Either that, or else it just recently appeared."

Mike took another sip of his tea before continuing. "Anyhow, since the Void on Earth is such a potentially contentious issue, without even considering the potential for political fallout with the Muslim Arabs and Jews in the area, we would like to send you to Mars and Titan to explore life on those planetary bodies and their relationship to the Voids and their energy signatures."

Jack was puzzled and taken aback by the blunt offer. He didn't know what to say so he stalled: "But I thought the scientific community agreed that there was no life on Titan? Lots of surface activity, yes, but the carbon isotope ratios were determined to be inconsistent with life."

"Maybe they are wrong." Mike shrugged his shoulders again for emphasis. "Those data and those conclusions were arrived at a long time ago by the Cassini-Huygens mission early in the 21st century. There were always some visionary scientists who suggested that any possible life there would be based on Titan's rich organic inventory, cryo-volcanic energy sources, and the presence of exotic solvents. Perhaps, after we discovered life on Mars and Venus and it was so similar to life on Earth, no one took it seriously anymore that some other form of life could exist there, based on a completely different type of biochemistry. We probably got even more Earth-centered than we were ever before. It is, in some way, quite arrogant of us to assume that any possible life on a place as exotic as Titan would follow the same isotopic fractionation rules as on Earth. Plus, we really don't know the environment on Titan very well. We have the same problems that we had during the Viking life detection experiments on Mars in the nineteen-seventies, when we didn't understand the Martian environment well enough to interpret the results of the probes."

Mike set down his cup and leaned forward in his chair to be closer to Jack. His eyes glittered with excitement and his voice softened in a conspiratorial tone. "If these detections are right and Zadak and his colleagues are right, then there is life on Titan, and we have the chance to discover a second genesis on Titan. There is no way to have a panspermia exchange by meteorites between the inner and outer Solar System. They are just too far apart and with Jupiter as a gravity sink in between, the chances are just too improbable to even consider. Your job will be to find that life, and the Void." He halted for a moment to consider his words.

"However, if we do find the same type of biochemistry as on Earth and Mars and Venus, well then, we have a good case of directed panspermia, but more in line with the concept proposed by DNA's discoverer Francis Crick in the last century. Maybe someone put them there together with the Voids…" Mike raised his eyebrows for emphasis and waited for Jack's reaction to the implications of his suggestion.

"Wait a moment," Jack started up, incredulous.

"Oh, I know," the administrator took back the initiative. "That sounds too far out. But your discovery is far out, too. And we're only just beginning to think about what all this might mean. Anyhow, whatever the philosophical implications, our idea is that you go with a crew that includes Zadak to Mars and Titan in about seven months."

"Seven months!" Jack set back a bit, dismayed. He hadn't wanted to spend that much time back on Earth. "What do I do till then?"

Mike Hang became serious again. "After your long space mission on Venus and your recent personal loss, we feel that you need some psychological and spiritual help."

Jack's was trying to keep his calm. *Spiritual help*? He wasn't about to let Mike Hang or anyone else ram their religion down his throat.

Mike saw that Jack was about to object and forcefully continued. "We'll send you to one of our associates, Monk Dahai, who has a retreat in the Chinese countryside, near Beijing. We want you to spend the next six months with him and…"

Jack interrupted loudly. "What the hell for?! I'm perfectly fine. I don't need that crap. I can do some sessions with a shrink if you like, but…"

Mike Hang was firm. "Yes, you do need it, Jack. That is a condition for your going on this mission. We need to make absolutely sure that you are psychologically and spiritually stable, especially on such an important mission, which could last a long time. And there is special training needed for this mission."

Damn Christian fundamentalists, Jack thought. How could they have one of them as NASA administrator? And how could this man just ram his own philosophy into a mission? Jack was appalled by the system. It was ironic that as science advanced more and more, the zealots appeared to get stronger and stronger. Maybe there was something cruel and cold about science, but he never felt it. Anyway, they could do in their churches whatever they wanted to do, but they should leave space exploration to the scientists. Jack sat in his chair, speechless.

Mike continued. "Look, Jack. See it like this. We need to make sure that you are completely fine."

"I *am* fine," Jack mumbled.

"But if you weren't then you would still protest just as vehemently. We wouldn't send you back up any sooner even if you had a flat psych test. We need at least half a year to put the mission together and finish the vessel. We have to modify and partially rebuild one of our Earth-Mars vessels for the much longer trip to Titan. The choice is yours, Jack. We would like you to command this mission, but if..."

"All right, all right," Jack interjected. "I'll do it."

"Very well," Hang replied as he rose from his chair. He opened the door and yelled into the hallway. "Send in Dahai!"

A few moments later Dahai stepped in. He doesn't look much like a monk, Jack thought. He was dressed in a white flannel shirt with loose pants and sandals. He was tall and had soft but firm brown eyes. He folded his hands in front of him, bowed to Jack, and just said, "Please, follow me."

Hang just smiled when Dahai and Jack stepped out of his office. Jack had the feeling he was being set up. They left the building and Dahai waved for a taxi. Dahai nodded in a perfectly friendly way toward Jack: they got into the taxi and he instructed the driver to take them to Washington National Airport, where they boarded the next plane to Beijing.

It was a long, silent flight. "Is there anything you still need to take care of before we start your training?" Dahai inquired as they were landing in Beijing.

"No, I took care of all my family matters after the funeral of my wife," Jack responded.

"You know that this is your last chance. Once the training has started there is very limited contact with the outside world. We do not have any telephones or computers in the lodging area, just in the conference room of the Complex, and those are only available to trainees in case of an emergency."

"I'm aware of that. I read the guidelines you gave me during the flight," Jack said flatly. He was still not happy about having to play this game and made no attempts to hide his skepticism.

"Very well, then," said Dahai, and returned to the silent repose he had shown during most of the flight.

Jack was not eager to start the so-called training at all, but as long as it provided him with the opportunity to command the expedition, he would endure it. And besides, there wasn't much else for him to do—no family left.

No life besides his career. The discovery had stirred quite a bit of interest, but most of the scientific establishment was conservative, and most of them were just asking questions about the discovery—questions he couldn't answer. Or they doubted his measurements, or worse, made fun of them. That was getting very frustrating. The only person he would have liked to talk to was Carmen, but she was literally millions of miles away. Besides, he didn't want their special connection to deteriorate over time with meaningless chit chat. There wasn't really anything he felt ready to share or to open up about, at least not via a relatively open, non-private digital transmission. Any conversations under those conditions would have come back to chit chat, so it was better to just forget about it.

The air was hot and humid when they left the airport. The taxi moved quickly through Beijing. Jack was amazed: this was one of the most technologically advanced and largest cities. Only a few decades ago, Beijing had been a chaotic mix of the modern and the developing worlds. They went to the western edge of the city where it quickly became very mountainous, passed remnants of the Great Wall, and finally arrived at the training center, which everyone referred to as "The Complex". A person in front of the main gate of the Complex greeted them with a bow. Dahai bowed back and they walked into the training center.

Dahai showed Jack his Spartan sleeping quarters. "You better get some rest, Jack. Tomorrow you start your training." Dahai bowed and left.

Jack was fairly tired from the trip and didn't need encouragement to seize some rest. He made himself as comfortable as possible on the hard bed, and went to sleep.

At 4 a.m. sharp a person who introduced himself as Lorenzo stepped into his room. Light was just barely visible through the window on this early summer morning.

"I brought you a robe and some sandals that should fit," Lorenzo said with a smile. "We all walk around here in this type of clothing. It is very comfortable during meditation. Please watch."

Then Lorenzo sat next to Jack and started his meditation. He sat down in the traditional yoga lotus position. At one point he extended his arms up, trying to reach the sky or heaven, Jack presumed, and was smiling. When he was finished Lorenzo said to Jack, "Master Lao will explain and teach you this meditation later. It is very powerful. You tap the energy from Heaven and Earth."

Jack wasn't sure how to respond. Tapping the energy of Heaven and Earth by sitting in an uncomfortable position and breathing deeply seemed a bit hokey and far fetched. So he remained silent as he got dressed. Then Lorenzo led him to a meeting room, where several other people were waiting. Lorenzo left with a bow. Dahai welcomed him and introduced the others to Jack.

"This is Dr. Chan, our psychologist, who will be responsible for your mental and physical health. This is Master Lao, your spiritual guide. This is Major Akinawa, who will attend to your physical fitness, and that is Sensei Hauser, who will train you in martial arts. And of course, I will spend quite a bit of time with you, as well. I will also be responsible for your overall training program. Additionally, in about three months time we will have a gentleman coming from the space agency, who will keep you updated on your mission, technology that you need to know, and so on. But we don't need to concern ourselves with these issues at the present time. Are there any questions?"

"No. No questions," Jack replied. He would be spending a great deal of time in the company of these people and wanted to have a healthy working relationship with them, even if he didn't really want to be involved in their philosophies. His own primary goal was to get back into space as soon as possible and this mission required him to train in a way he had never before experienced. He would give it his all, even if he was skeptical of the practical benefits of the training. In a way he was excited to be experiencing something new. That's what exploration and science were at the basic level.

"Good," said Dahai.

All the others left with a bow toward him: only Major Akinawa and Dahai remained.

"You will start your training program with Major Akinawa and will afterwards meet Master Lao for spiritual training," Dahai instructed him. "Akinawa, are you ready?"

"Hai," the Asian man said. "Follow me." His voice was soft, but stern. It spoke of strength, confidence and discipline.

Jack followed Akinawa and they started with a brisk run on a trail around the facility. They changed the dynamics with intermittent sprints. After fifteen minutes Jack felt exhausted. But, of course, giving up was no option. After an hour of running, they continued with exercises, 121 push-ups and 121 sit-ups. What was the magic about 121, Jack wondered. Jack had to take breaks: Akinawa waited patiently, but made it clear with his demeanor that he expected one hundred percent commitment only. Shortly after 6 a.m., the fully exhausted Jack was led by Akinawa to Master Lao.

Lao sat in a lotus position and waved toward Jack.

"Let's take a walk in the garden."

Jack was wondering why they would take a walk in the garden when it was raining outside, but nevertheless welcomed this idea very much as it allowed him to rest. He followed Lao through a moon gate into a courtyard with a beautiful Chinese garden. Limestone rocks, plants and a large pond were interwoven in a serene and tranquil setting. In the middle of the garden was a hill with limestone rocks. On its summit was a pavilion stretching the edges of its roof like wings to the sky. It was raining lightly and rain water cascaded down from the drip tiles at the eaves like a crystal-beaded curtain. Lao led on the difficult, winding path to the pavilion. A sweet scent of flowers filled the air.

"This path pleases your mind," Lao noted. "You are focused on the path. Therefore you have no time to worry about other things. It also teaches you that the path, the process, is the goal, not the result. The result will come automatically if your process is right."

Lao sat down in the middle of the Pavilion and gave Jack a sign to do so as well. Lao started meditating, while Jack sat patiently next to him. The rain slowed to a trickle and stopped, and the sun peeked through the clouds once in a while. The air was loaded with an earthy smell. The pond next to them reflected the pavilion, rocks, and plants. They sat there for quite a while: eventually, Lao addressed Jack. "Time has no meaning in this place. The garden strives for perfect harmony, for balance between hard and soft. Look at the pine tree growing next to the limestone, the pine is soft and shows that even though life is so fragile it has the power to grow in tough and treacherous conditions. Look at the bamboo. It represents quiet resilience; it bends in a storm, but never breaks. Look at the willow; it portrays feminine grace and beauty. How empty would life be without it! Look at the winter-flowering plum tree, it shows renewal; a new start; a new beginning."

"Please, breath the air; the scent," he continued. Jack sniffed at the flowers.

"No, no, much deeper! See?" Master Lao took a deep breath. Jack tried to imitate him.

"Not through your chest. Breath deeper! Through your abdomen, through your whole body. Imagine yourself breathing in oxygen for every single cell in your body. Every cell needs to be nurtured and nourished."

Jack tried hard to do so. "Is this better?" he asked.

Lao indicated so with a small nod.

"Now, inhale with joy. Joy of the wonderful creation that is around you." Jack tried to do so.

Master Lao shook his head, "Is that all the joy that is left inside you?"

Jack tried to do better.

Again, Lao shook his head.

"I think you became too much of a scientist. Analyzing, analyzing, but without the ability to sense the wonder in it anymore. Try to think back to your childhood. You were born with this ability."

But the thoughts of the past did not fill Jack with joy. Master Lao wrinkled his forehead and said, "I believe we have a lot of work to do."

Then they went into a room and the Master instructed Jack. "Move your legs shoulder-width apart, put your knees to the front with the weight on the front part of your feet, push your stomach forward, spine straight, adjust your head slightly backwards, and relax your back muscles!"

Then he had to stretch out his arms with his palms facing each other. Master Lao took a stick and laid it across Jack's wrists. Jack wobbled a bit. Then Master Lao gently corrected his posture, and said to him "Stay like this until I return," and he walked away.

After five minutes Jack was wobbling and sweating profoundly from exertion as his muscles strained to continue holding the stance. He was getting frustrated by his body's weakness. This should be easy, he told himself, and gritted his teeth in determination. He forced himself to continue; to ignore the pain. What would he have given to be outside, running with Akinawa instead! A few minutes later Master Lao stepped back into the room, corrected his posture, told him to continue, and left again. Jack went through phases where he thought he was strong and could hold on for much longer and phases where he was ready to give up and just couldn't stand it anymore.

Another fifteen minutes later and Master Lao returned. It felt to Jack like hours. Master Lao said gently, "That is enough for today. Eventually, you will do this for several hours at a time."

Jack released his mental hold on his body parts, exhaled forcefully, and resisted the urge to sit down right there on the spot.

"Your muscles can support you only for several minutes, but your Chi can support you for much longer. The longer you can do these exercises, the greater your Chi will be."

"Chi?" Jack asked.

"It is difficult to explain. There is no analogue in the English language or in Western culture. The best description is probably *breath*, or *life energy*. It represents the key essence of what you are. Quite literally it is air. Air encompasses you everywhere and lets you breathe and live. Our whole being is based on that, on this energy field. Our body is only a material manifestation of this energy field."

That didn't make much sense to Jack's scientifically trained mind. Jack was used to think conventionally, but he chose not to question further at the moment. Then, Master Lao continued with breathing exercises, this time in a sitting position.

"Remember to inhale as deeply as you can," he repeated over and over. "Breathing is life. Every cell of your body needs oxygen, and this is the only way to get it to them."

Master Lao instructed Jack in the meditation technique, which Lorenzo had demonstrated in his room earlier that morning. Jack mastered the technical aspects very quickly. Master Lao called the technique *Jesus sitting*. "But now remember to breath! Combine! Inhale with your tongue at the roof of your mouth, pressing against your palate; inhale deeply along your spine, and release your breath through the front of your body."

They went through the exercise several times with Jack still struggling. Mostly, Jack had problems coordinating movements and breath. Master Lao finally said that it was time for breakfast. He took Jack to Lorenzo's room, and Lorenzo led him to the breakfast table.

"How are you feeling?" Lorenzo asked.

"Very tired." And then he added, "This is the time I usually get out of bed."

Lorenzo gave him a big smile. It felt good to have Lorenzo around.

"I wonder," Jack said, "when I was standing in the posture and at various times afterwards, I felt something…." He hesitated a moment as he searched for the right words. "…a strange sensation between the palms of my hands. There was an additional odd thing. The further I put my hands apart, the stronger the feeling became."

"This is called lao-gong. The interior of your palms are a natural energy outlet. This is also the area where you sweat first. Your experience is quite normal. What it means in the form of a scientific explanation, I don't know. Maybe it is your biological electromagnetic field or some other energy field. I really don't know. But this is only the start of what you will experience as you progress."

"I will take you to your martial arts instructor after breakfast," Lorenzo continued.

"Why do I have to learn martial arts?" Jack asked and joked, "I don't think that I'm going to have a fight with space aliens any time soon."

"Oh, there is much more to learn from martial arts than fighting. I think they have assigned Hauser as your instructor, so that you will realize you needn't be Asian to be good with Chi."

Sensei Hauser awaited him on the mat. Hauser bowed to Jack and he returned the bow.

"Come face me," he said. "Now, strike me!"

Jack hesitated. Hauser nodded for encouragement. Jack finally lashed out after Hauser. The master just blocked Jack's fist with his open hand.

"No, no. Strike me as if you really mean it. With the intention to hurt me!"

Jack did so. Hauser just stepped to the side, touched him slightly at his shoulder, and Jack fell in front of Hauser's feet.

"One more time!" Hauser said. "And again."

Jack was amazed. Hauser barely touched him and yet he was falling toward his feet like a dust particle being sucked into a vacuum cleaner.

"What you should learn from our little exercise is that aggression is not a useful answer to aggression. If you are relaxed and calm, then you have the power. Not muscle strength, I talk about power like the eye of the hurricane. The eye is totally calm, yet it is the center of power. Strike me again!"

And Jack was falling faster than he realized.

"Second message," Hauser continued. "While you were attacking me in a straight line, I moved you in a spiral that took you off balance. The spiral is the most powerful shape in nature. Think about a hurricane, a tornado, or DNA. Notice that I use your energy in my favor. When you attack, I connect with you and become your third leg when you execute the technique. And then I take it away, gently, and you fall. You do it now!"

They practiced just the one technique over and over for hours, but Jack still was clumsy. He made a multitude of mistakes, such as forcing the technique with his muscles, which only produced resistance in his partner, or he lost control of his balance and center of gravity. At other times, when focusing, he was only focused with his mind rather with his whole body. Or he had coordination problems. His mind was overwhelmed with the intricacies of performing a seemingly simple task. It should have been easy, but the timing, coordination and focus necessary to flawlessly execute the task were elusive and the process was beginning to frustrate him.

When they separated, Hauser emphasized, "In order to understand, you have to realize that there are more than just physical aspects to the technique. You are a spiritual being, like it or not, know it or not. If you become really good, then you will do the technique spiritually, not physically. The faster you understand and accept this truth, the quicker you will advance."

## 2.2 The Witch Cage

Then Dahai appeared. "I will take you to your next training element," he said. Jack wanted to faint. His body had experienced more exercise already today than it had in the last few weeks altogether. He wasn't sure if he could take much more.

"Don't be alarmed," Dahai continued. "The next training element will be tough. But we do not intend to harm you, although it was used in the middle ages of Europe as a torture tool."

Jack stopped in his tracks, unsure about exactly what the hell he was getting himself into.

Dahai stopped and turned towards him. "Everything here has an important purpose and we only have a limited time to get you ready for your upcoming tasks."

Jack's fatigue made him cynical. "So, this won't be as pleasant as my previous training elements?" He was painfully aware of his exhausted body and aching bones. They generated thoughts of doubt and rebellion.

Dahai saw how physically tired Jack was and replied. "Oh, don't worry about your body. Your body will rest, but your mind and spirit will be challenged. You have to fight your inner demons." Jack felt uneasy but was intrigued at the same time.

Dahai led him to a room totally painted in black and fixed him into a kind of harness, the only thing in the room.

"We call it the witch cage or truth squat. You will face your inner self, you don't have any choice. It works with sensory deprivation. You will be strapped and positioned like this for about an hour and then you will be slowly rotated for the same time. They used this device on accused witches in the middle ages to make them confess, but we employ it here for a much better cause. You learn to be at peace with yourself and with the energy surrounding you, which is for many people the hardest thing to do. We also adopted it because it can nicely simulate space travel when you are by yourself for long periods of time."

Dahai tightened the harness and left the room, turning off the lights and leaving Jack in total darkness and silence.

Thoughts bombarded Jack's mind as his senses desperately searched for something to hold onto. There seemed to be no up and down, no fixed point whatsoever. Only himself, only his mind! His mind seemed to be in a total flux with past, present, and future merging. Images appeared from the past, at times overwhelming. He tried desperately to keep calm. The intensity of the emotions increased and Jack struggled to keep his composure. When the harness finally rotated he was not sure whether to cry or grab onto the little hope he still possessed that he would manage to survive with his sanity intact and still ready for the other half of the "torture". He became dizzy when blood poured into his head while he was hanging upside down. Then his body appeared to adjust, but the emotional pressures and strains grew more intense by the moment. The demons of his past haunted him! His thoughts shifted to and fro, out of his conscious control. At times he became panicked and frantic with the thought that he was losing his mind. At other times he just wanted to let go and cry like a baby.

Finally, the promised end arrived. The torture was over. Lorenzo came in, straightened him up, and freed him from the harness. "Are you all right?" Lorenzo asked, concerned.

"I'm not sure." Jack was visibly shaken and flushed from his ordeal.

"We take it for granted don't we—that we are grounded? And if we are not, that's deeply troubling for our consciousness to interpret and react to."

"Were you in that thing as well?" asked Jack

"Sure, until I finally learned to be at complete peace with myself."

"I was afraid of that." Jack grimaced and straightened himself to regain his composure. "I guess that will not be the last time that I will be in the witch cage."

Lorenzo just smiled and led him out of the room and to the dining area for a late lunch.

"You know what I just found out?" Lorenzo said in a conspiring voice.

"No, what?" said Jack, much more concerned with his brutalized body than with anything Lorenzo had to say.

"Dahai used to be a high ranking officer in the U.S. Army. A colonel or a general."

"How did he become a monk?" asked Jack, now slightly interested.

"Apparently, he became enlightened and quit the military. He then built up the Complex as a meditation and training center."

"Hmmm, that would also explain his relationship to Mike Hang."

After lunch Lorenzo led Jack to Dr. Chan, the psychologist. Jack couldn't believe how drained he was, physically, mentally, emotionally, and the day was not even half over. Having just eaten made him even more tired.

Dr. Chan asked Jack to sit down in an armchair. Probably, the most comfortable piece of furniture in the whole training center, Jack thought.

"Would you like to talk about yourself? Or perhaps some of the experiences you had today, and how they relate to your past?" Dr. Chan asked gently.

"I don't know. I'm not sure." Jack rubbed his forehead intensely. "I thought I handled it all well with my wife's death, my experiences from my youth, but after the last few hours in that cage I'm not sure anymore. Not sure about anything, really. But maybe I only need some time."

"Time does not resolve anything, at least not by itself. You have to deal with the demons of your past. Well, perhaps you will feel more certain about things tomorrow. Perhaps we should just use our time to get acquainted with each other." For nearly an hour Chan told his life story and about various traumas and hardships that he had experienced while growing up in a poor eastern province of India. When he was finished he curtly stood and then called Akinawa in to take Jack to more exercise.

Jack sighed, but steeled himself—he would survive this intact! He dreaded another workout, but anything would be better then the thrice cursed witch cage.

This time the exercise looked to be even more rigorous than in the morning and Jack wished that he had managed to stall things by talking to Chan, if only to cut down the duration of the exercise and extend the time spent sitting in that comfortable armchair.

Jack closed off the part of his mind that felt pain and discomfort. He willed himself to be in some other place, imagined himself in the middle of huge fields of ripe strawberries, and let the pain and fatigue pass through his body. The exercises passed by in a foggy haze. Jack was only modestly aware of his own presence. Time had no meaning during this numb period, but finally it was over and Jack was given a small reprieve.

After the exercising Jack had an hour of private time to shower and rest, before he went off to another martial arts training session with Hauser. Since

Jack was all out of energy, Hauser did iron body training with him—basically, Hauser punched him and Jack had to endure the pain. Jack almost laughed as the session started, but caught himself. His fatigue was making him delirious—perhaps the walls that protected his sanity were being tested? When Hauser was done with him, Jack went with Lorenzo to dinner. He met two other trainees, who were in various stages of their training program, but Jack was far too exhausted to hold a conversation or even remember their names and the meal passed by in a dreamlike haze. After Lorenzo and Jack had finished eating, Jack was taken to a meditation chamber and led through a series of breathing and relaxation techniques, guided by Lorenzo. When it was over, Jack returned to his room and dropped dead-tired into bed. It was only 9 p.m., but he had never felt so tired in his entire life. Finally, the first day at the Complex had ended.

The next morning started at sunrise—4:30 a.m. Jack's body hurt so much that at first he didn't think he would be able to even get out of bed. Akinawa picked him up for the physical exercise routine. Once he started the exercising, the pain receded. Damn little Japanese guy, Jack thought, a true energizer bunny. Then it was time for martial arts with Hauser again, who tried to drill into his brain the need to relax when executing martial arts techniques rather than to muscle them. Another major theme was to use his whole body rather than upper body strength only. Hauser commented that he apparently hadn't succeeded at softening up Jack's muscles enough during the iron body training the day before, and that they would have to repeat it this afternoon or evening. After feeling completely exhausted, he met with Master Lao to improve his posture and his breathing. The various stances were challenging, perhaps more so than yesterday due to his exhaustion and various body aches. But it felt good to focus and try to ignore his physical discomfort. In the early afternoon he had to be a victim of the dreadful witch cage again, but he was more mentally prepared this time, or at least he thought he was.

Master Lao and Chan were secretly monitoring him through a hidden window while he hung upside down.

"Do you think he is ready to open up?" Chan asked Lao.

"Wait until tomorrow," Lao suggested: "I see him crying. His emotions are starting to come to the surface and he's beginning to deal with them, but there is still too much resistance. Mold, but not break. That is the art. I would recommend letting him hang an hour longer today, and afterwards send him to get a physical examination. You can have your next session with him tomorrow."

"So be it." Chan replied.

During the physical, Jack had to supply urine, blood, and bone marrow samples. He hated those tests; one unfortunate sample result, and all the effort could be for nothing—and he would not be able to lead the space mission.

But the mission was all he was living for. Still, the physical exam was better than exercising and a thousand times better than that damned witch chair.

After the physical it was again Hauser's turn, continuing to emphasize the total commitment needed to execute and master a martial art. "You need to be ready to die, otherwise you are not a hundred percent committed. What is the problem? The end-result of your life is death anyway, right?"

"The problem is that I don't want to die yet," Jack responded.

"That is not your call," Hauser corrected him. "Either way, you cannot be totally committed without it. In order to be at your best you have to be willing to offer your life. To detach from your life!"

Jack was flying and falling in so many directions that he completely lost his orientation several times and would fall over again before he could stand up. Jack got sick and tired of one torture after another. Was it really worth it? But he continued. He had his doubts but he fought them down every time. There would be no second chance for him. If he dropped out, where would he go? There was no other future.

The third day started the same, but this time the witch cage was even worse. How he hated it! In the following session with Chan he couldn't help nearly breaking down: tears flooded his eyes while he moaned and lamented that he was a total failure; that he hadn't been able to make his wife happy; that he failed to commit to having a family; that there was no one out there who gave a damn about him. He submerged himself in self pity. Chan asked him why he felt that way, and Jack elaborated on his life and the many times he had been hurt and disappointed. The feeling of being lost and confused during his younger years, and that he still had doubts and feelings of not having a clear idea what he wanted from life. That perhaps he had joined the space program to escape from life and to avoid having to deal with people.

Chan didn't say much but just let Jack go on and on, for hours. At the end he asked a few targeted questions, assured him that the Complex was his new family, and that they cared about him. Then he sent Jack to his room.

By the fourth day, he felt much better; at least mentally, although Akinawa seemed to be especially set to drive him to complete physical exhaustion. Hauser actually praised him by acknowledging that he'd already become more relaxed in his martial arts. The witch cage was still grueling but not as bad as the day before—but nonetheless, it was the training element he despised the most, even though it helped him become detached. The fourth day was less focused on physical rigor and Jack had long conversations about his past and present with both Master Lao and Dr. Chan. They drilled into the issues Jack had revealed the day before in his session with Chan, and Jack felt naked, exposed and embarrassed for having to share his emotions, fears and insecurities so openly with others.

On the 5th day Jack had a long meeting with Dahai. He was discouraged with his progress and believed that the instructors viewed him as weak. "I take it that I haven't been performing very well?" he asked Dahai.

"On the contrary. You are making progress. As long as you are here, we will succeed together. Remember, you are always free to leave. That would be a shame, though. I think you show great potential. Of course, we always have backup plans for the mission if you aren't able to complete your training. But Mike Hang and I genuinely believe that you are made out of the material that we need for the mission."

Jack welcomed the encouragement but wasn't as certain as Dahai. "I feel my progress is excruciatingly slow."

"This kind of training takes time," Dahai reassured him.

The break Jack had hoped for on Saturday didn't appear. But in the evening Master Lao came into his room and told him that Sunday would be time off.

Jack's tired bones were aching and even his mind appeared to be hurting. "So, I really get one day off?" Jack was astonished. It sounded almost too good to be true. Like a random fantasy he never expected would really ever happen.

"Sure." Master Lao smiled. "Even God took off the seventh day for rest. There has to be a time for things to settle in. Time for you to reflect on what you've accomplished and how much you have grown. And of course, to realize how far you still have to go!"

Although he wanted to sleep much longer, Jack was already awake by 6 a.m. so he got up and strolled into the garden, sniffed a flower here and there, and enjoyed the fresh morning air.

Master Lao was in the garden, too, and greeted him in his friendly manner. "I see you enjoy the garden."

"Yes, I do," said Jack truthfully.

"That's very good."

"I'm sorry to be so difficult and disappointing at times." Jack did not feel comfortable admitting his weaknesses or shortcomings, but he wanted Master Lao to believe in him.

"Hush, hush. I never have understood why you westerners are so concerned about what other people think about you. Detach yourself from such concerns! Things are just the way they are. That's all."

Jack was amazed. The man could read him like a book. "I'm glad you think like this."

The two sat in silence for a while, just absorbing the beauty and balance of the garden until Jack felt it was time to move on. He rose, politely bowed and silently left the garden to return to his room and listened to the radio for a while. The last week had appeared like an eternity to him, but nothing much had happened in the outside world. He quickly got bored of listening to the radio and shut it off. There were no digital displays in his room and nothing really else to do, so he went for a hike in the neighboring forest, read a bit of a book that he found in the library of the Complex, and recorded a message to Carmen. He thought that he had something valuable to tell about his training experiences and some of the insights that he felt he had gained, and he wanted to share them with her. But before sending off his message he added a note that he would not be reachable for the next six months, which was only partially true. Maybe, he was simply afraid that she would respond? Or was he more afraid that she would not? Anyhow, he had to learn to be centered in himself, not around someone else. In the evening he enjoyed a game of chess with Lorenzo, and then a volleyball game with three other trainees. He fell asleep that evening more content than he had been in a very long time.

Then, the much dreaded Monday came and with it the program started all over again. However, there were moments when he started to believe that his body, mind, and spirit were becoming stronger, more in unity, and ready to face whatever would come. And challenges were still coming. Like the time when Master Lao had him stand in one of the postures for five hours. Or the day when he was rotating in the witch cage for a whole day. His sessions remained one to one, just him and the instructor, for the first month, but then slowly shifted to incorporate other trainees. His martial arts training, for example, was now conducted together with two other trainees, one big, bulky guy named Olaf, and a cute Chinese girl who was able to flip him on his back in an instant if he didn't watch out. Later, the physical training and the sessions with Chan were at times conducted with others. The witch cage session was not scheduled every day anymore, but occurred more and more infrequently. However, whenever scheduled, it lasted for at least half a day. He developed some comradeship with the other trainees, especially during the free Sundays.

## 2.3  The Crew of Deep Explorer

At the end of the third month, Dahai approached him and told him that they would have to start preparing him for certain technical and command details for his upcoming mission. A specialist from the Office of Space Exploration was scheduled to arrive next week, and start training him. Jack looked forward to that. He was anxious to move on to the next part of his life; namely the mission and getting back into space.

The mission planning person, Robert "Bob" Watson, arrived to bring Jack up to date on the mission details. Bob was a typical NASA bureaucrat, Jack thought, but Bob was also top-notch in engineering and design issues. The assembly of the spacecraft was progressing according to schedule. It had a state-of-the-art propulsion system—a combination of ion propulsion and nuclear-chemical propulsion. Its maximum speed would be 0.12% of light velocity. The habitation ring was built to rotate around the main spacecraft. The spacecraft would carry a crew of six to nine people for extended periods. Bob elaborated that the trip to Mars would take about twenty-six days and to Titan another ten weeks. Most of the time would be needed for acceleration and braking. Bob got very excited and animated while presenting the technical specifications. He also proudly described the newly developed quantum entanglement chip, nicknamed the "QE chip", which would for the first time ever allow instantaneous communication between their vessel and Earth headquarters eliminating the annoying time delay problems experienced in earlier missions that used electromagnetic signals or the more primitive precursors of the QE chip for communication. It seemed on the surface like a violation of Einsteinian physics, but in fact it was not.

"You should be able to inspect the spacecraft in two to three months and we should be ready for launch in four months," Bob informed Jack.

"Wonderful!"

"In a month or so you will also have your crew. We are planning on a crew of seven, including you. You'll travel first to Mars to study the Martian Void. You've been allotted a time slot of two months for this expedition. Your main job, however, will be to lead the exploration of Titan, precisely locate the Void there, analyze it, and find any life that is associated with it."

"Sounds good to me," Jack replied confidently.

In addition to the mission planning and tech training, Jack's rigorous training at the Complex continued. Most of the time now, though, his training was conducted with fellow trainees, except for one-on-one meetings with Master Lao, Dr. Chan, and Dahai at regular intervals. His martial arts training now included weapons, and fighting multiple opponents. Akinawa trained him for extreme endurance under physical stress. The most enjoyable part was the long philosophical conversations that he had with Master Lao, when he was not practicing breathing or meditation exercises.

In the middle of the fourth month at the Complex, Jack's crew arrived. The crew included Zadak, the mission scientist, who studied the Void data from Venus, Earth, and Mars. They seemed to get off to a good start and mutual understanding, right away. Zadak was an excellent rationally-driven analytical thinker. Jack also enjoyed Zadak's distaste for politicians. The crafty scientist made it very quickly known to everyone that he considered politicians mere parasites living off the hard work of the working population and that they are responsible for pushing humankind from one disaster to the next. Zadak admitted freely that his convictions derived in large part from his cultural background, claiming that politicians and leaders had been hampering the development of his ethnic group, the Serbs, for more than five hundred years.

Then there was his assistant scientist Adana Birsh, who was as excited as a little child about the mission and revered Zadak like a devoted fan worshiping a sports hero. She had grown up in the Boston area in a wealthy, privileged family and was Zadak's opposite in nearly everything. Jack wondered how this odd scientific couple had ever found one another. Next, there was Etienne Garibaldi, the chief engineer, and Garibaldi's assistant Pablo Weniger. Garibaldi was somewhat older, in his late 50s and slightly balding, while Pablo was in his mid-30s with a promising future career. Garibaldi was not a typical engineer, but had quite a bit of science background including a Master's degree in plant pathology. It was only late in life that Garibaldi, as he said, had found his true calling in engineering. Pablo grew up in a Los Angeles suburb, eldest son of a German sailor and a Mexican mother. His mother had been an illegal immigrant. Pablo joined the military early, and after four years in the U.S. Air Force he went to school at some of the top engineering schools and finished with high honors. Their pilot was Lana El'Barais, a woman of mixed Arabic and black descent from Nigeria. Lana was about Jack's age, in her early forties, but looked younger. At first she appeared to be stubborn and headstrong to Jack—possibly a problem in the future. But then he re-considered and realized that a pilot should probably feel strongly about her judgment when flying a spacecraft. Only the physician's position was still to be filled and the right person had not been determined yet.

The crew went through the same training as Jack, but not quite as rigorous. In addition to being continually trained himself, Jack also participated now in mission planning and in evaluating his crew members' performance. Joe and Kitahari, two astronauts in their mid-twenties, joined the training of his crew, as possible replacement science personnel. Joseph Palati, called Joe, was from Italy and excelled at being funny on nearly every occasion. He was very skilled in putting a grin on everyone's face. His other expertise was planetary environments. Kitahari, on the other hand, was of Japanese descent but had been raised in San Francisco. She was definitely a more serious type. Very quiet, but deep and determined in her goals, which included her religious beliefs. Her previous training was predominantly in biochemistry but included a variety of other talents.

The anticipated crew, including Joe and Kitahari, conducted many training activities together, such as the early 10 km run with Akinawa, which became a part of their routine. They didn't have much social contact though, because the training activities fully consumed their days. Jack was concerned about this and confided to Lao that after three weeks he still didn't know much more about his crewmates than their names and expertise. Master Lao assured him that this would not be a problem, because they would have enough time on *Deep Explorer* to get acquainted with each other. For now, the individual training had to be the focus. Once Jack passed by the room with the witch cage and heard Kitahari screaming through the walls. He felt sorry for her as he recalled his own experiences, but there was nothing he could do. She had to overcome it and resolve her issues, just as he had.

Nearly six months passed since Jack had started his training at the Complex, and Dahai, Bob Watson, Master Lao, and Jack were evaluating the crew.

"Jack, you should know that Adana told me that she will drop out of the mission. Her father is very sick and she feels that she needs to take care of him. She is leaving tomorrow," Dahai said.

"Yes, she told me about her problem. That's too bad. Who do we replace her with?"

"That's really your call. Joe scores higher in nearly all training exercises, but Kitahari may be better for the homogeneity of the crew. Otherwise Lana would be the only woman left."

"Kitahari is also strong in her beliefs, which gave her strength through all the training exercises and may be useful during the upcoming mission," Lao added.

"Well, true." Dahai responded. "She came in as the weakest person and we wondered if she could make it. But she hung in and actually has one of the steepest learning curves."

Jack thought it through for a moment. Then he said, "Since it is my choice, I want both, Joe and Kitahari."

Dahai looked at Lao and Bob, and said, "Sure, why not? The spacecraft is designed for up to nine crew members. Let's do it."

## 2.4 Mission Briefing

Finally, the day of the mission briefing arrived. It was one month before the scheduled launch. The astronauts, including Jack, waited anxiously in the conference center for the briefing to start. Mike Hang came, with Graig Fuller, a NASA engineer at his side, and a scientist from the Russian Space Center -Igor Brosewich- whom Jack did not know. Dahai and Master Lao were present as well.

Mike started the briefing. "Thank you for being here, all of you. It is great to see our finest. We are making great progress in building your ship, but we still need to complete some upgrades due to a possible extension of the mission."

"Graig," and he turned to his engineer, "please give the crew an update on the ship and what it can do."

Graig, a bulky man in his fifties with gray hair, got up and flashed a technical drawing on the screen. "Basically it is a big bird, 47.8 m long, with the cockpit and the command room right up front and a habitation ring rotating around it. You will live in the habitation ring most of the time. You have artificial gravity in the living ring. It is suited for a crew of seven to nine. Each crew member has a private compartment, and there is a communal exercise room. In the main section of the ship, including the cockpit, you will have no gravity, so you will need to wear magnetic boots that make you stick to the floor. But the entire vessel is pressurized, so you won't need your spacesuits inside. The ship has various compartments that are separated by air locks. If the hull is breached somewhere, the sensors on the airlocks will register the drop in pressure, and the doors will seal automatically. There is a communication center in the main body of the ship to stay in touch with Earth via a QE chip. Also on board are a fully equipped lab and an imaging compartment with the latest remote sensing technology. Below the cockpit we've installed a powerful laser strong enough to melt a rock on the surface of Titan even if you are thousands of kilometers away in orbit. The huge generators in the back of the ship, here and here, next to the engineering compartment, provide the power."

"What is the huge laser for? Fighting space aliens?" Zadak interrupted.

Graig looked to Mike. Mike simply said, "We don't know. You are very far out there. We have to consider all options."

Zadak got upset. "That's ridiculous. You want to play Star Wars out there? That's typical government. Ah, I get it. You want to blow up the Void! You find something you don't understand and you want to blow it up."

Mike Hang raised his hands cautiously to calm Zadak down and stem his tirade. "We are not there yet and hopefully won't get there," Mike replied. "I admit there are some political factions in various governments that would like to do that, but as long as I'm in charge that won't happen. And this is a NASA-led mission," he added.

Zadak calmed down. Mike Hang nodded to Graig to continue.

"All right, the spacecraft has two shuttles attached to its belly. One is the usual shuttle with which you are all familiar; Discovery class. The other is what we call an Ice Scooper. It can eject fifty-kilo projectiles at phenomenal speed, via a miniaturized rail gun. If you target one at the planetary surface the impact will eject a lot of material into the atmosphere or orbit, and then the mouth of the shuttle opens and can scoop up that material, which can then be analyzed in the shuttle or in the main spacecraft.

"I don't believe this," grumbled Zadak.

"Both laser and scooper are excellent tools for scientific analysis," Craig continued, ignoring Zadak's comments.

"Or weapons of mass destruction," Zadak commented.

"Only if they are in the wrong hands. But we trust that they are in the right hands with you. That's why we subjected you to the rigorous training," Mike added.

Jack rubbed his forehead. He didn't want to have to deal with malcontents or political paranoiacs amongst the crew and hoped Zadak wouldn't be a problem. But even Jack felt ambivalent about these high-power tools that could easily be used as weapons. "All right, what is the science or the mission objective specifically that calls for the use of these tools?"

"Okay, let's talk about this. Igor, please." Mike replied and turned to the Russian scientist.

Igor was humorless and to the point. "Objective 1 is going to Mars to analyze the Void that has been found in a side canyon of Valles Marineris. You will land at the Mars Base Station, fly one of the shuttles through the canyon and scout for a good landing location. You will approach the object and analyze it with various scanners and with a variety of non-invasive methods. Most of all we would like to know how it compares to the Voids on Earth and Venus. Total time for the Mars investigation is one and a half months. Any questions about Objective 1?"

The crew and guests remained silent so Igor continued. "I thought so. Mars is straightforward. Objective 2 is to fly to Titan. In contrast to the Voids on the inner terrestrial planets, the energy pulses from the Titan Void are different. We don't know what to expect there. Your task is to locate the Void. That should be relatively easy. We have provided your spacecraft with the most sophisticated and sensitive scanners available. It shouldn't be a problem to home in on the energy signal. Since we believe that the Voids are in some way related to life, we want you to find that life on Titan, if it is there. That may be a bit complicated. That life may be completely different from what you expect. If it is on the surface, it may be living in liquid ethane or methane, or in ammonia-water puddles covered up by water-ice. Or, it may be in a deep water-ammonia reservoir or even in underground water. If that is the case, you will have a cryobot on board which can melt itself through the subsurface into a subsurface lake or into ground ammonia-water."

"What about planetary protection measures and protocols?" Jack asked. "We don't want to risk contamination of either Titan or ourselves, and by extension Earth."

"We will decide this with the Planetary Protection Council on a case-by-case basis and inform you at the appropriate time," Igor replied.

"All right then," Garibaldi said and got up.

"I'm not finished yet," Igor said. "There may be an Objective 3."

"*May* be?" Joe emphasized.

"Yes, may be. We picked something up coming out of the Kuiper belt. We first noticed that object two years ago, but didn't think much of it. We estimate it to be roughly the size of Saturn's moons Enceladus or Tethys. We have just recently confirmed that it also emits an energy signature very similar to a Void, and it may actually have a Void of its own."

"You expect us to fly to the Kuiper belt and check it out?" Lana asked.

"No." Igor continued. "It actually just passed Uranus and continues to move towards us. It appears to be on a direct course toward Earth." Jack straightened in his chair: that was alarming news. The whole crew was reacting in a similar manner.

Zadak rose in his chair and pointed an accusatory finger. "That's why you've mounted the laser on the ship. You don't have any idea what's out there and you want us to take it out."

Igor responded in his humorless monotone. "Please allow me to downplay your paranoia, Sir. Our measuring sensitivities at this range are not very good and if we are off in our projections just a little bit, it can easily be heading somewhere else. Towards Venus, or directly into the Sun."

Mike Hang took a breath and said, "Obviously, this mission is very important. We are scientists and explorers, ladies and gentlemen. Not militaristic xenophobes. If the calculations of Igor and his colleagues are correct, then you will be near Saturn at the same time as this object. If so, we need to learn as much as possible about this object before it comes close to the inner Solar System. I received this request from our planetary protection officer. This may be all about nothing—it most likely is—just a Kuiper Belt object that has a strange trajectory and will probably be flung out of the Solar System by Jupiter or Saturn." Mike paused and then said, "Any questions?"

Garibaldi raised his hand.

"You said that this object had an energy signature similar to the Voids. Is it a Void? If so, how strong is its energy signature?"

"The data are not clear. We could be wrong, but it appears to change in strength, sometimes barely registering, at other times as strong or stronger than Earth's Void."

"Oh," Mike added, "…Since we are still doing some upgrades to your ship, we decided that you will stay here a bit longer while you get your detailed tech briefs and training in your specific tasks. The launch should be no more than six weeks from now. Dahai was gracious enough to allow us to use his facility a bit longer. Given the importance of this mission, I'm sure you understand." By the way, your vessel will be named "*Deep Explorer.*"

After the presentation Jack approached Mike Hang. "I did some thinking," Jack said earnestly. "I was wondering; if I should become incapacitated or die during the mission, who would take my place?"

"You will have another member that you know pretty well and who is ideally suited for that position," Mike replied.

"Who?" asked Jack, curious as to whom Mike was referring to.

"Dr. Chan. We finally decided to fill the physician's position with him. He is more a psychologist than MD, but we've trained him in other medical issues as well in the last couple of months."

Jack was stunned. "I would think that I should have a say on this?"

"You object? Sure, you should have a say. We'd like him to be on the mission, because he should be very valuable, especially in difficult situations. But if you choose to have someone else as a second in command during the mission, we would support that. Perhaps, Garibaldi, based on seniority. Just based on rank, without any formal appointment, the Chief Medical Officer would be second in line. Would that be acceptable to you?"

"Yes. I guess Chan would be an asset. He knows the entire crew very well and vice versa."

Privately, Jack wasn't too happy with Hang's choice. Probably it was because he felt vulnerable toward the person to whom he confided his whole life, including all his failures. That was also not ideal if he had to rely on issuing authoritarian decisions in difficult situations. But then, Chan would truly be the most suitable Chief Medical Officer for the mission, and once in space he would just promote Garibaldi to be second in command.

Zadak met Jack in the hallway some time later and exchanged a few words with him. "What do you think about our mission?"

"I truly am not sure. I guess I'll have to sort it out," Jack replied.

"They didn't tell you more? You are the commander!"

"Not a word. Only one thing is clear in my mind—unexplored territory. A very interesting mission is awaiting us."

"Come on Jack. You heard them in there. The unknown object with bio-signs on an intercept course with Earth. Then there's the laser and the rail gun. We don't need that crap to do 'science'. I don't trust them one minute. They're not telling us everything and you know it. Aren't you worried about their intentions?"

Jack didn't want to have to address Zadak's opinions, even if they mirrored his own, but they were incendiary and Zadak's aggressiveness on the issue wasn't a quality he thought would be an asset on the mission. Jack didn't like wild cards and it crossed his mind that it still wasn't too late to replace Zadak, if necessary. He needed team players on this mission, not independents. He straightened his posture and tried to sound more confident than he felt. "I think they just want to be prepared for any type of circumstances. Don't be paranoid."

"Don't be paranoid? Are you joking?" Zadak was on the verge of exploding. And the last thing Jack wanted was for the episode to get back to Mike Hang. It wouldn't reflect well on the mission or on Jack's leadership ability.

Jack put on the most serious face he could muster and looked Zadak square in the eyes. "Zadak, I'm only going to tell you this once. Either you get a hold of your fears or you'll find yourself staying here. I don't have any idea what we are going to find out there, but consider this—this is one of the most important missions humanity has ever launched into space. Who do you want to be out there? You? Or someone else, perhaps one of the military guys?"

Jack glared at him for a moment longer and then broke away and moved on to his other duties.

The next day Jack had a meeting with Dahai. "Please come in and have a seat," said Dahai. "This mission becomes more important every day. We need to increase your and your crew's training to have all of you optimally prepared for the mission. The mission might be longer than we originally planned."

"Master Lao mentioned that," Jack replied.

"Did he mention also that we expect you to join our spiritual service on Sunday?" Dahai continued calmly.

"No!" Jack said. "Isn't that a personal issue? I don't like to go to church."

"If it is only about you, yes, then it is. However, you have a crew that is also looking forward to your spiritual leadership."

"What do you mean?"

"This is obviously a dangerous mission. Every mission is, but this one even more so. If one of your crew members dies, you have to give that person a funeral. That has always been the duty of any ship's Captain. Even worse, you might find yourself in despair, in very bad circumstances. What are you going to do then?"

"I will do what needs to be done," Jack replied evenly.

"Exactly, and for that we need you to be there for the spiritual service. Let's face it. Humankind has always looked for something greater than itself in the universe. There always has been the need for a spiritual being. Your crewmates need to see you there, so they have trust and faith in you. It is not for your sake."

"I didn't know that my crewmates were such church-goers," Jack snapped.

"Oh, not all are. Zadak is definitely not, but then he is only the science officer and not the Captain, like you. You should be there so they see you. It is an image of solidarity that will be very important to them."

"All right, I'll be there. Anything else?"

"No, not really." Dahai concluded. "But while you are there keep your heart and spirit open. You may actually enjoy it. I think you've changed quite a bit in the last six months."

Jack just shook his head and repressed a smile as he walked away.

As promised, Jack went to the 'spiritual service' and greeted Garibaldi, Pablo, Joe, and Kitahari, who were visibly happy to see him. He wondered whether this was a trend—that most of the younger scientists were more spiritually inclined nowadays—or whether this was a new selection criterion instigated by the new NASA administration under Mike Hang. As the service progressed he found that he liked the music at least, if nothing else. It provided nicely peaceful surroundings, but he was puzzled how especially Pablo and Kitahari were so much into swaying bodily to the rhythm. Must be nice to be so free and to release like that, he thought as he stood like a rock in the oscillating crowd. Well, that wasn't his thing. But the many friendly faces did make him feel better and he agreed with Dahai that his presence was beneficial to some of his crewmates, and maybe, who knows, to the mission as a whole.

The next few weeks of training were the hardest yet. The staff of the Complex exposed them to intense training continuously and sometimes it appeared that the staff was actually out to kill them by physical and mental exhaustion. It was amazing that nobody got seriously hurt. Most of the physical exercises and mental and spiritual stress exercises were now conducted in the afternoon and evenings, while the tech briefings were in the morning. Zadak had taken Jack's warning and calmed sufficiently and Jack was much more confident that he would be a valuable and productive member of the crew.

## 2.5 Winds of Destiny

About a week before their scheduled liftoff Jack flew to Washington, DC, to meet with the NASA administrator. He felt excited that it was almost time for the mission launch. The spaceship was completely assembled and waiting at the Earth Orbital Station. The engineers were currently triple-checking all functions, modules and backup systems. Jack's only fear was that something would go wrong at the last minute. His nervousness only grew as each day brought them all closer to the reality of launch. Nonetheless, he felt elated and confident when he walked through the corridors of NASA headquarters. Mike Hang's secretary let him into the office, but asked him to wait a little until Mike returned from some other meeting. While he wandered around the room, he examined the pictures on the wall which portrayed the history of space exploration underlining the biggest successes starting from the first lunar Apollo landing to the establishment of a permanent base on Mars. Soon NASA would celebrate the 100th anniversary of the first Apollo landing. After a decades-long lull after the lunar landings, during which hardly anything happened space exploration finally kicked into high gear and mankind had established permanent human bases on the Moon and Mars. Also, the Venus Orbital Station had been manned a while ago. Jack admired the artist's conception of the orbital station around Venus. He got some of the details wrong, but did pretty good, Jack thought. Even these baby steps of space exploration would probably not have happened if they had not discovered life in the atmosphere of Venus and the Martian shallow subsurface. These discoveries finally provided the political and popular momentum for space exploration to really kick off. Of course, this had its costs, too. There had been no mission to Titan since the Cassini-Huygens mission more than fifty years ago, and only one mission to the Jovian moons—which crash-landed on Ganymede due to a navigational error. Thus, their mission to Mars and Titan was really a new bold step.

And *he* would command that mission. The mission was also the first trial to test the newly designed propulsion system on an extended mission. Perhaps, with the new propulsion system, exploration of the Solar System would finally move more quickly. Instantaneous communications via the QE Chip would help immensely, too.

Mike Hang entered his office then, which catapulted Jack out of his thoughts. "Hi Jack, how are you doing?" Mike greeted him warmly.

"It looks like we are nearly set," Jack said.

Mike Hang nodded in agreement. "Yes, it has been a long process, but the moment we've worked for is knocking on our door."

"Let's hope we don't have a major glitch at the last minute."

"There shouldn't be any. We've employed the best team for the job and did not shy away from any expenses to make the mission a success."

"Let's hope you hired the right crew for the mission as well."

"I like to believe so," Mike said with a smile.

"On the flight to D.C., I was wondering who would have led the mission if I hadn't made it through the training. Or if I had just left the Complex?"

Mike Hang sipped on his coffee. "We had two other trainees for the captain's position. Let's just say that one couldn't take the training and left, the other one unfortunately lost his sanity. But he is recovering right now. And if everything else would have failed, Chan would always have been an option. But you were always our first choice."

Jack was astounded. In some way he admired how frank the administrator was, but the news of the wash-outs was alarming. He knew the training was intense but he was surprised and concerned to hear that the administrator endorsed imposing such a potentially harmful regimen on his trainees. It doesn't fit the profile of a religious person either, Jack thought. "I hope you will be just as frank with me when we are on the mission."

"You bet." Mike reassured him. "I don't like being any other way."

"All right, you called me in here for a meeting," said Jack. "Anything new regarding the mission?"

"Yes, we have confirmation that Objective 3 is part of your mission. It seems that the Kuiper Belt object with its Void, the KBO, is real. This is a very serious concern now."

"Was its trajectory confirmed?"

"Yes, it is currently on the shortest trajectory to Earth. It'll pass quite close to Titan given its distance and velocity. I don't have to tell you how unlikely it is that this is simply by chance."

"How long until KBO reaches Earth?"

"Eleven months and twenty-five days, give or take a few. You have to intercept it and figure out what it is. You and your team are our forward sensor. After your interception, we only have six to seven months left until it reaches Earth—assuming its velocity remains constant."

"You don't think that this KBO is inhabited or steered by intelligent aliens, do you?" Jack said half jokingly.

Mike remained serious. "I don't know what to think. All possibilities are on the table and have to be considered. Whatever it is we need *you* to figure it out. That may be much more important than Objectives 1 and 2. You need to intercept that object to determine what it is."

"Will do." Jack said. "I understand the mission quite clearly!"

"One more thing. I've got some bad news and I think you should be one of the first to know. There has been an accident with the crew on the Venus Orbital Station."

Jack turned pale and sat back.

"Janine Ludmilla, the person in charge, tinkered around and tried to interact with the Void. We really don't know yet what she tried to do. Some say she had special instructions from the European or Russian Space Agency. Either way, the result was a large energy burst that killed Ludmilla and badly injured Mejilla.

"How bad was Carmen injured?" asked Jack, jumping out of his seat. He hadn't had much time to think of Carmen with the hectic schedule of training and mission planning. But he was forced to admit that he had feelings for her, as strange as that sounded considering they only knew each other for a few days.

"Well, we don't know exactly. But it appears that half her face was burned pretty seriously. She made it back from the surface to the station, though, but looked pretty badly burned in the telecon. She seems to be in a state of near-complete agony. Worse still, there is not much in the way of medication stored in the orbital station to reduce the pain. We asked her to take the shuttle and return directly to Earth, put the station on automatic, and we would take care of the replacement crew. But frankly, I don't know whether she will make it, and if she survives at all, whether she can keep her sanity during the trip home."

Jack was shocked. Not again, he thought. Another loss. He felt drawn to Carmen even though they hadn't had the time to ever get really close. Perhaps he should have established contact to her after arriving on Earth. Or perhaps it was better not to, as he had decided. He remembered Lao's words not to attach to Earthly things and to be prepared to give up any attachment like this. Only in this way it is possible to live, he said. Only if you truly accept death will you obtain the ultimate freedom to make the best out of the time given to you.

For a moment he thought that there might be something wrong with him, another person close to him having an accident. But then his analytical mind dismissed this notion as ridiculous. And after all, Carmen, as he knew her, had something special in her, some kind of natural strength and positive attitude toward life. She would come through no matter how badly the cards were stacked against her. He forced himself to believe that and was confident in her future.

"I knew her a bit. She is tough. I believe she will make it," Jack replied after a pause.

"I hope so."

As Jack left the administrator's office, Carmen's image popped into his head over and over again. He imagined how badly she was burned and how alone she was, and in so much misery. He wished he could be there for her. Tension rose inside him. Finally, he went to a telecom port in the NASA building and got in touch with Fred, the communications officer.

"Hi, Fred." Jack said.

"Wow, Jack, it is you! I have heard you will command the mission to Titan. Is that true?"

"Yes, it is. But I have to ask you for a favor in another matter."

"Anything. What can I do for you?"

"Mike Hang told me about the accident on the Venus Orbital Station. I used the last fifteen minutes to compose a little email message. Would it be okay if I send it to you and you forward it to Carmen Mejilla?"

"I can do that, but I don't know whether she is in any condition to check incoming stuff."

"I know, but I need to try."

"All right, consider it done!"

"Thank you, Fred."

Jack executed the 'sending email' command on the telecom. The message read:

*No matter the storm, its enduring intensity*
*No matter how the deck of cards is stacked against you*
*You will prevail*

*Prevail like a rose that is peeking through the ice on a cold, frosty morning*
*With the promise to bloom once again*
*Even brighter than ever before.*
*You will make it. I believe in you.*

*Jack*

Jack was far too upset to think straight. His emotions were in turmoil and he was almost frantic with concern. Then he took a deep breath, sat down and meditated as Master Lao and Lorenzo had taught him. He searched for the greater balance; for his inner calm. He ignored the NASA civil servants running by. At some point he felt a deep and comforting sensation. After a while he was confident that he had mastered himself and his emotions. He stood up. He felt relieved and calm. He even felt that Carmen would be fine. The inner nervousness was gone. He pushed away any remaining gloomy thoughts, took a taxi, and arrived at the shuttle port about half an hour late.

The East Coast Shuttle Launch Facility near Blacksburg, Virginia, accommodated Jack's delayed arrival. He boarded the shuttle to the Earth Orbital Station, or EOS, to which *Deep Explorer* was docked. The pilot was already waiting. The EOS was huge, and equipped with cranes and assembly facilities to avoid costly launches from Earth's surface. The old International Space Station made up its core area, but so much construction had been added that the components of the former space station were difficult to find. EOS could be seen from Earth with the unaided eye as a mosquito-like structure. Jack felt the strong jolt and g-forces as the shuttle lifted off. He still felt like he was in somewhat of a trance. But he had to get out of it. He had to focus on the mission and its future. This had to be his objective. As the shuttle approached EOS he could make out *Deep Explorer*, which looked like an appendix stuck to the station. A few minutes later he docked. His senior engineer was already waiting to give him a tour.

"What do you think of your new ship?" Jack asked Garibaldi.

"Very, very impressive. It is very large! Most of the space is taken up by the propulsion system and the huge generators for the laser, though," he said with a smirk.

"The dreams of an engineer, I presume. Not much people-space, eh?"

"Not much for a crew of eight. There is the cockpit and a compartment right behind the cockpit in front of the ship, the command and communication room in the center of the ship, the engineering compartment near the thruster engines, the lab, the cargo bay, and of course the habitation ring. The thing I like the most is that you don't need to wear a spacesuit anywhere inside the ship. It feels like you're in your own living room at home, except for the magnetic boots."

Jack and Garibaldi entered *Deep Explorer* from the EOS docking ramp, donned magnetic boots, and walked through the entire ship checking out its main functions and operations. Jack fought down the queasiness of weightlessness. It was going to take a while to get used to switching back and forth between zero-G and this artificial gravity. There were medications for it, but he hated popping pills. It was better to let his body adjust naturally.

"It is strange not to have the shuttles in the cargo bay," Jack commented.

"Yes, the cargo bay doors were actually designed too small, intentionally. They couldn't make them bigger because of the huge propulsion system and the volume needed for fuel storage. So they had to improvise and anchor the two shuttles outside, on the belly."

Jack just remembered Objective 3 of their mission and asked Garibaldi "What's our top speed?"

Garibaldi grinned. "Faster than that Kuiper Belt object they are talking about. I'm sure we can get up to 0.1 % of light velocity easily, but for that we need a lot of acceleration time. The most amazing thing really is the maneuvering thrusters; how quickly we can make turns and change directions. The acceleration time to 50 km per second is quite impressive. Of course, we need a lot of propellant for that."

"I'm glad that you are happy with our ship," Jack concluded and thought that a happy engineer was a good sign things were going the right way.

Jack joined Garibaldi for dinner and then took the next shuttle back to Earth.

Two days until the scheduled departure. Jack visited Master Lao for the last time. "I wish you could come with us. I would feel much better if you could join us."

"Oh, space exploration is for young folks," Lao replied. "I'm sure you will do fine on your own. You have been trained well. You are well prepared to deal with any unforeseen challenges that cross your path."

"Mike Hang seemed to be concerned about the Kuiper Belt object heading our way."

"So am I," agreed Lao. "I think we all are. There seems to be something deep, dark and forbidding about it. Difficult to grasp."

Jack was surprised by what the old man told him. Jack was a man of science, but the training he experienced at the Complex made him hesitant to dismiss these kinds of feelings out of hand, especially when it came from Master Lao.

"Can you be more specific?" Jack tried to probe deeper.

Lao shook his head. "Not really. The specifics are your job to find out. Just be aware. Don't dismiss the spiritual side of whatever you will encounter out there. Remember what sets us apart from the robots. They have much higher computing abilities, but we have four billion years of evolution to our advantage. Embrace your humanity. Trust your instincts."

"I will try to remember that," Jack said, admiring Master Lao for his wisdom and insights.

"Here. I have a little present for you." Lao reached into the folds of his robe and pulled forth an envelope. "Open it when you are leaving orbit," he instructed with a smile.

Jack accepted the gift reverently and bowed to Lao. "Thank you."

They walked for a while in the garden, mostly silently, inhaling the scent of the flowers. Jack was intensely aware of how silence can be so much more meaningful than idle conversation.

The next day Jack was heading to the New Mexico Space Harbor in the U.S. for launch preparations. His plan was to arrive in El Paso that evening and take a cab to the spaceport. Jack arrived just before sunset when the bare mountains were glowing vibrant red. Jack felt little dust particles impacting his face whenever a wind gust would lift up some of the loose soil from the ground. It already looks like Mars here, he thought as he drove north through the Chihuahuan desert. The waters of the Rio Grande glimmered in the fading light in an otherwise monotonous landscape overgrown by mesquite, sage brush and prickly pear cactus. Jack took the time to appreciate the lonely beauty. Despite its desolate appearance the desert was crawling with life. Most of it just chose not to reveal itself to casual observers. The comparison to Mars, and Venus and Titan too for that matter, was rather apt. Superficial appearances could be quite deceiving.

When the last vestiges of light reflecting on the patchy mountain top snow retreated, his thoughts focused again on the upcoming mission and the challenges he would face. Finally he made it to the Space Harbor. He had his bags sent to his room and then immediately made his way to the launch facility, which was enveloped in a loud buzz of activity and construction noise.

He saw Joe standing at a com terminal flipping through a binder. "Excited?" Jack shouted above the clamor.

"You bet," he smiled widely, his face brimming with energy. "This has been my dream since I got out of diapers."

Jack laughed and clapped him on the shoulder. "Yep. I'm ready to get the heck out of here and into deep space, too. Where are the others?"

"As far as I know they're messing around at Rabbit's Lounge—that's the recreation room at this facility. I'll be heading over there shortly. Just one last hurrah to remind us of home till we get back." Joe was as enthusiastic as ever.

Jack nodded and left Joe. He didn't like the fact that the crew was playing when there were still mission duties to attend to, but he wasn't going to begrudge them a little enjoyment and relaxation.

He only had to ask two people how to get to Rabbit's, where he found the rest of the crew nursing beers and mingling with about twenty other local officers, scientists and engineers. Zadak, Kitahari, and Lana were sitting together. When they saw Jack, they waved him over. Jack put on a smile and strode over to them. Well, he thought, if you can't beat them, join them. He was quickly introduced to Lana's husband and forced to engage in mission small talk for far longer than he felt comfortable doing so. After an hour of mingling he excused himself and retreated to his room.

The next morning he met the remaining crew members, except the engineers Garibaldi and Pablo, who had already gone on board. After a wonderful breakfast, they boarded the shuttle to the Earth Orbital Station without any fanfare or press. The shuttle arrived at the station twenty minutes later and after exchanging a few pleasantries with the station commander, they quickly boarded their own vessel, where Garibaldi reported that all systems were ready and functioning.

The view down to Earth was stunning. It was a picture of precious beauty, which anyone would remember the rest of their life. It really did seem a fragile thing from this vantage, Jack thought. "Take a good look folks; this will be the last view for a long time."

No one said anything.

Lana took the controls and Jack sat next to her in the co-pilot's seat. Garibaldi and Pablo were in the engineering compartment to monitor the engines during launch, while the rest of the crew were in the command room near the center of *Deep Explorer*.

After an extensive and lengthy pre-flight checklist Jack contacted Mission Control through the telecom link. "Mission Control, all systems are ready to go. We are ready to launch."

"Permission granted."

The countdown started. Jack got that tingling in his gut, like whenever something new and exciting is imminent. His thoughts drifted to his late wife and then to Carmen. As the countdown from mission control continued through the speakers, Jack pulled out the envelope that he had received from Master Lao and opened it. He carefully unfolded the letter inside and read it.

*I have studied many times*
*The marble which was chiseled for me –*
*A boat with a furled sail at rest in harbor.*
*In truth it pictures not my destination*
*But my life.*
*For love was offered to me, and I shrank from disillusionment;*
*Sorrow knocked on my door, but I was afraid*

*Ambition called to me, but I dreaded the chances.*
*Yet all the while I hungered for meaning in my life*
*And now I know that we must lift the sail*
*And catch the winds of destiny*
*Wherever they drive the boat.*
*To put meaning in one's life may end in madness,*
*But life without meaning is the torture*
*Of restlessness and vague desire –*
*It is a boat longing for the sea and yet afraid.*

*Edgar Lee Masters*

"What is that?" Lana asked.

"Oh." Jack responded ruefully. "Just a good journey wish from an old friend, a wish for a great adventure to come."

"Three…two…one…"

The engines ignited and *Deep Explorer* left Earth's orbit on its way to Mars.


# 3  Mars

## 3.1  Martian Base Station

"Our mission is truly a frontier mission," Jack said when they had their first daily crew meeting in the engineering compartment on day one of the trip. "We should be on Mars in less than four weeks. What is the status in engineering?"

"The primary engines are fully operational," Garibaldi responded. "But we still have some problems adjusting the thrusters. The aft starboard maneuvering thruster is only functioning at forty-six percent capacity and some of the other thrusters exhibit a similar power shortfall. I suspect the injectors are becoming clogged. If I enrich the fuel mixture a bit, that should do the trick. Also, some of the auxiliary systems don't check out precisely right, but nothing major."

"Don't worry about it Garibaldi. This ship was thrown together in a hurry and I'm sure we'll continue finding bugs for weeks. So long as these problems don't kill somebody or delay us, I wouldn't fret about it. I'm sure we'll find and fix most of them by the time we reach Mars. Dr. Chan, what is the medical status of the crew?"

"All in good health and spirits."

"Anything new in science?" Jack asked.

"We have a better location fix on the Martian Void," Zadak said. "It is definitely in the bottom of that unnamed Valles Marineris side canyon, in Candor

Chasma. Mars Base Station reported that its signal is about two to three times as strong as the one from Venus."

"Anything new from Titan or beyond?"

"No, not that I have been informed of."

"Okay, any other issues or concerns?" Jack asked the entire group.

"There seems to be a problem with one of the recyclers," Joe answered.

"What's the problem?" Garibaldi inquired.

"The UDD is leaking so I'd advise us to get it fixed soon," Joe added, referring to the Urine Disposal Device.

"I'll check it out," Pablo volunteered.

"Very well then," Jack said. "Attend to your duties, until tomorrow morning, same place, same time!"

Jack toured the ship afterwards. His goal was to become familiar with each system and each operational unit. There hadn't been sufficient time to do so during training, and it bothered him that he wasn't as intimately knowledgeable of his ship as he had been on other missions. He was proud and enthusiastic about leading this mission, but in many ways he still wasn't comfortable with it. Maybe it was the fact that it was hastily thrown together. Maybe it was partly because he didn't feel he had been allowed to interact or bind with the crew enough. Maybe it was the fact the *he* was in command and responsible for the mission, the ship and the crew. Or maybe it was just the mission itself and these mysterious Voids. Whatever they were, they didn't seem "natural" to him.

As Jack toured the ship he took the time to speak with each crew member one-on-one. Chan is right, he thought, everyone is in good spirits. All were happy to be off the ground and eager to begin the mission. Hopefully this would last for a long time. At the end of the tour he went to the cockpit where Lana was sitting. He sat down for a while and stared into space.

"Isn't it funny," Lana said, interrupting the silence. "When I was a kid I was so scared of big spaces. When I was swimming in a lake, I thought I would sink to its bottom or get sucked into the vast volume. And now I'm in space."

"Yes, it is strange that we usually get exposed to that which we are afraid of the most." Jack continued to stare into space for a while and then went to his room for meditation. Jack had enjoyed the meditation exercises at the Complex at the end of his stay and intended to continue them on a daily basis. It calmed him and it also seemed that he could draw strength from them. He would probably need all the strength he could get.

At the meeting the following day Jack asked every crew member to spend two additional hours a day to become familiar with systems for which he or she was not an expert. That was most difficult for Lana as she had to give every crew member except Pablo a short course in how to fly *Deep Explorer* and the shuttles. The days went by quickly. The crew was working hard becoming familiar with their vessel, preparing for their Mars deployment, and at the various training sessions. No one appeared to miss the gruesome physical, mental, and spiritual training at the Complex. They all joked about their experiences as 'hell on Earth', but most of the crew members, including Jack, maintained their routine of meditation.

The happiest person on board was Garibaldi. He enthusiastically went about his engineering chores with the single-minded devotion of a man who had found his true passion. He was ecstatic about their vessel and what it could do. Needless to say, Garibaldi was very proud when the modifications he had made allowed them to arrive at Mars a whole day early.

Everyone was eager to arrive on Mars and their anticipation grew steadily as each day brought them closer. Being able to put their feet back on solid ground was a simple luxury that they had taken for granted. The rotating habitat ring helped a lot in simulating Earth's gravity and making the trip more comfortable. It would also reduce the bone and muscle loss during their mission to nearly undetectable levels. But nevertheless, there was nothing like standing on some firm ground, a planet, and sinking your own body into the planet's gravity well.

Their approach to Mars was routine. The navigation computer automatically oriented the ship and plotted an approach vector. Lana manually piloted the vessel into a stationary orbit over the Martian Base Station. The strain of the insertion caused a few warnings and glitches to show up on the ship's auto-diagnostics program. Jack decided to leave Garibaldi and Lana on board to oversee the ironing out and fixing of any remaining glitches in the ship's system operations. He boarded the shuttle with the others, and piloted it to the Martian Base Station. It was quite a bit different from flying on Venus or Earth. The thinner atmosphere took a few moments to adjust to, but he had no problems homing in on the station's signal beacon.

"Ain't nothing but a smooth ride," Joe grinned as they entered into the station's main hanger bay. The hanger could be closed off from the outside but it wasn't pressurized, so the crew had to put on their enviro-gear to enter the station proper. They had to pass through the decontamination chamber first before they could enter and be officially welcomed onto the station. It took twice as long as usual because the decon chamber could only treat five at a time and they were six altogether.

The station commander, a tall, slender upright Brit, greeted them in the hallway. "Great to see you, Sir!" he said to Jack as he grasped his hand in a warm welcome. "I've heard so much about you!"

"Thank you. It's great to be here. Mortimer Upham, I presume?" Although Jack had been in contact with the Martian Base Station, he hadn't been able to directly talk with the commander until now.

"Indeed, Jack. Just call me Morty. We are very informal here in the base station at the frontier of humankind. Please allow me to show you the station," and he led them into the command room.

"This is our command and meeting room," Morty continued. "It is the center of the station with the habitation zones extending from the center like spokes of a wheel. Our main power supply is next to the center, as are our assembly and engineering facilities. Do you need anything from us for your expedition?"

"Thank you, but I believe we have all the supplies we need in our shuttle. We would appreciate one of your gas hoppers and could use some of your local expertise."

"Ah yes, no problem. Our hoppers are a treat. We have finally worked the bugs out of the first models. Our engineers are very proud that they were able to optimally adjust the pump to an ideal setting for drawing in carbon dioxide from the atmosphere, storing it in a liquid form, and then sending it through a preheated pellet bed turning it into hot rocket exhaust to produce lots of thrust. You will get the prime model with the large solar panels and the ground rover. It will all be set up for you."

"Great. I really appreciate that. How would you recommend getting to the Void?"

"Oh, I thought about that already," Morty illuminated a light table and then continued. "I believe the best way is if you land on this small mesa in a neighboring side canyon, go down this natural ramp with the rover, climb down the cliff with a rope, just about a 15 m drop, and then work yourself through the boulders from this ancient landslide to the bottom of the canyon. Once there, you have a hike of about two and a half kilometers, and you should be right there.

"I assume there is no way that we can go there with the gas hopper directly?" Jack inquired.

"I'm afraid not. These canyon walls are just too unstable. The hopper could cause a landslide too easily. It would be an unnecessary risk to say the least. We can go through the details later. But let's have dinner first. We have a special treat and I would like for you to meet my people. We currently have a permanent crew of six, including myself."

Jack had already studied the station's crew manifest and their profiles while he was on *Deep Explorer*. He already knew one of them—Daniel Perera. When Danny saw Jack in the dining facility he threw up his arms and shouted, "Mr. Bubbles!" Then he ran over and grasped his hand excitedly. "Welcome to the station, old man. You became famous, I heard—Jack, the Discoverer of the Voids!"

Jack just smiled and shook his head. "I just knew you were going to be trouble for me," he teased. "And I'm only a few years older than you, kid." It was good to see Danny again and he was glad that his unrestrained, exuberant personality hadn't dimmed in the least.

"Bubbles?" Dr. Chan asked. "Did he just call you 'Bubbles'?"

Jack turned to the doctor and gave him a half serious glare. "Don't bother asking, because I'm never going to tell. And Danny promised me long ago that he wouldn't ever tell another living soul how I got that nickname. Unfortunately I didn't make him promise never to say it aloud again." Jack loathed the nickname, earned in college during a drunken celebration. Danny wasn't even present at the time but apparently he had heard the story from someone else. Jack had no idea the embarrassing event would come back to haunt him here on Mars.

"How do you know Danny?" Kitahari inquired.

"Oh, Danny and I went to college together, but Danny was just a freshman when I was a senior so we didn't socialize much. We didn't meet again until we were at NASA headquarters when I was just beginning my training for the Venus mission. Danny came and found me, and asked me and my wife out to dinner. I didn't even recognize that little punk kid from college," Jack said with a smile, "…but apparently I made a big impression on him and inspired him to become an astronaut, too. At NASA headquarters he trained to become a back-up mission specialist for the Martian Base Station. I'm glad to see that he made it."

Dinner was an elaborate and special affair. In addition to meeting and mingling with Morty's crew they were treated to an incredible meal of real food. The station maintained an impressive and diverse greenhouse. Tomatoes, carrots, turnips, green beans, radishes, peppers, and corn were available in tempting abundance. They had to eat dehydrated beef and chicken, but by space standards they were living like royalty.

The social interaction was friendly and warm. They talked about family and Earth. The residents told stories, some serious and some funny, about their experiences on Mars, while Jack's crew told their colleagues about the harsh training at the Complex. Morty and his crew all agreed that they were glad they never had to endure that misery.

But business was on the menu, too. They all discussed the mission that Jack was leading and what he knew about the Voids. They worked out the details of the expedition to the Martian Void in regards to equipment and logistics. Marina Caruso, one of Morty's more experienced pilots, had volunteered to fly the crew to the site in the hopper. But she would have to leave them there and return to the station. Jack was hesitant to place his crew in such isolation. "What if an accident occurred and someone is seriously hurt? How would an emergency extraction be carried out?" he protested.

Morty insisted that it was a logistical necessity. "I need the hopper for the functioning of the base station and even if I didn't, well, I don't want to leave it exposed to the Martian elements for that extended period."

"Martian dust has a nasty habit of getting into everything and breaking down mechanical components," added Marina. "Furthermore, piloting a hopper on Mars is more difficult than it seems and I wouldn't trust it in the hands of an inexperienced pilot. Winds are chaotic and can get really nasty without any warning."

"Also, your drop-off point is so far from the base camp that even if there were an emergency, your crew members would still have a heck of a long walk just to return to the hopper. Marina could fly to any position she's needed much faster from the station," Morty concluded.

Jack had no choice but to agree. His primary concern was for his crew. The Void was obviously dangerous, but Morty was right about the logistics in the event of an emergency. Either way, this was Morty's station and his equipment, and therefore his decision.

Kitahara was interested why they called their planes 'gas hoppers' and Marina happily enlightened her about the history of the Martian planes. The first flying vehicles actually just hopped over the ground, several kilometers per jump.

When dinner finished and the festivities gave way to duties, Mortimer asked Jack to join him for a walk in the station. "What do you think about that Kuiper Belt object, the KBO, that's coming our way?" he asked Jack.

"I really don't know. I haven't had time to give it much thought really." It wasn't a very honest answer. Jack had lain awake many nights contemplating this strange object heading toward Earth, but he still didn't have any idea what to make of it. "We will deal with one problem at a time, Morty. First we have to do a job here, then on Titan, and after that we will deal with the incoming planetoid. Or whatever it is, if it is still an issue then."

"That's probably a wise attitude. I'm just glad that it is you who is going out there and checking it out."

"I hope you feel the same way in a couple of months."

"Oh, I'm sure. After your encounter, we might be the next ones who have to deal with the KBO."

"Well, I'll try to give you a good heads up before it comes close to you," Jack reassured him. Then he returned to his guest quarters on the Martian Base Station for the night, as did his comrades. Since the station could house twelve they actually got to spend the night in real beds.

## 3.2  Candor Chasma

The sun rose over Mars and gave the atmosphere a red-pinkish hue. The exploration crew assembled near the gas hopper and went through the supply and equipment checklist. By 0700 Mars Standard Time they lifted off to Valles Marineris and the suspected location of the Void. The Martian landscape was incredibly beautiful in all its desolate glory. The crew was glued to the observation ports to soak in as much as they could. Craters, canyons, and lonely mesas decorated the landscape in shades of black, red and rust. While flying through the vast canyon system they admired the steep cliffs and imagined the huge catastrophic outbursts of water that had so long ago created the lower parts of Valles Marineris. It wasn't all that different from Earth really.

Jack smiled ruefully. There was still a big push at home to initiate a terraforming project on Mars. It was gaining governmental support from many fronts; mostly from the wealthy and influential members of society that wanted an off-world experience. Jack doubted whether most of them understood what they were really asking. Any change in the Martian environment, or introduction of foreign flora and fauna, would likely be the kiss of death for the life already existing on Mars. Moral issues aside, the cost and scope of the project would incredibly drain a world economy already struggling to keep the bulging population clothed and fed. Add that to the fact that it would take centuries of focused work to terra-form a planet that could never truly be earthlike. No, he thought, Mars is just fine the way it is.

The ship's sensors indicated that they were nearing the Void. Jack asked Marina to fly over the Void before attempting a landing on the mesa. It wouldn't hurt for him to get a visual perspective of the route they would be traversing.

They reached the location from which the energy signature of the Void was emanating. They could record the energy signature easily, 2.7 times as strong as the one from Venus, but orders of magnitudes smaller than the one from Earth. The pulsating pattern was slightly different, but they would have to do an in-depth analysis to gain more insights.

"There," Marina said, pointing to a series of large ripple like structures in the canyon. "It's down there."

Jack followed her outstretched arm, but on the surface he did not see any obvious hint to the presence of the Void. It was likely quite a bit underground. Marina had the gas hopper hovering over the identified location, but there was no way to land it safely: the steep ripples were studded with large angular boulders. So they flew onwards, to the mesa that Morty had suggested.

The gas hopper set down at the pre-planned landing site; a flat gravel strewn area with small crescent-shaped dunes locked in their slow march across Mars. In no time at all the crew put on their enviro-gear and began unloading equipment. Pablo squeezed into the driver's seat of the small rover and charged the power sequencer. When the green signal light blinked on he engaged the hydro-drive engine and drove the rover out of the belly of the gas hopper. Jack and his crew loaded the gear onto the vehicle.

"Everyone make sure that you have all your oxygen tanks, hook-ups, ropes, and your parts of the base camp setup," he instructed as they boarded the small rover. In a moment they were off and bumping along the Martian surface. They held onto the railing while the rover made slow progress over the rough terrain. It reminded Jack of the carts used in mines during the old days, the ones you could still ride in some places as a special tourist attraction. Or even of one of the old "covered wagons" from the American pioneering era. When they finally reached the cliff, everyone was more than ready to get their feet back on solid ground. Everyone knew their routine and duties. They jumped off the rover and attached ropes to rocks at the cliff-edges for the climb down.

Zadak and Joe climbed down first. Then the rest of the team lowered the base camp equipment down to them before Kitahari and Pablo, and then Chan and Jack descended. When they arrived at the bottom, they distributed the base camp equipment to the crew, which slowed their progress on the ground. In addition, the terrain became even tougher, with boulders and large rocks chaotically jumbled at the bottom of the canyon. In single file they went through the rocky terrain. On their way they noticed wet patches of greenish-reddish material from which gas bubbles gurgled up. Jack pointed to the bubbles and commented through the radio, "Metabolic activity. That's methane gas, generated by Martian life."

"That's all that is left from one global Martian biosphere? That's pretty sad," Chan commented.

"That may well be so. But it is still a form of alien life, which is exciting by itself."

"There is more than that," Zadak noted. "In some of the caves around here Morty and his crew found microbial communities that form slimes. Yet, I agree, there should be much more. If the energy signal is nearly three times stronger than the one from the Venusian Void, and if our hypothesis is correct that the strength of the signal is directly related to planetary biomass, then we know only the tip of the iceberg. There must be a lot of life underground of which we know nothing."

"I doubt that it is so straightforward that you can extrapolate the global biomass so easily from the field strength of the Voids, and the type of life from its pulsating pattern," Kitahara questioned Zadak's hypothesis.

"Maybe. Maybe not, but so far those predictions seem to be right on target, at least for Venus and Earth," Zadak emphasized to support his idea.

"Well, either way, that's why we came to Mars—to figure it out. We will work on an all-encompassing hypothesis to describe and explain the Voids," Jack noted. "And Titan will then be the test of our hypothesis."

After several arduous hours they finally reached the presumed location of the Void. It had taken a bit longer than expected. Pablo had slipped at one point and taken a nasty tumble off a low boulder ridge. He had banged his knee pretty hard on a rock, and although he had assured Jack that he was fine, he hobbled along with a limp and a grimace. The flashlight he was carrying got smashed and so was one of the hand scanners.

They unloaded their gear and quickly set about erecting the base camp. The base camp included a tent that could sleep six, a power generator, and an oxygen generator. The tent had the shape of an oversized soda can that was pumped up with oxygenated air, so the occupants would not have to wear their suits and helmets inside. After they set up the base camp and stored all other equipment and gear in its place, the day neared its end and the sun vanished below the horizon. It became very cold very quickly and Jack opened a heat canister to warm up the inside of the tent. They discussed strategy for tomorrow.

"We have to be real careful with the drilling. Our analysis from the sensors on the gas hopper indicates that the Void is not as deeply buried as we thought," Zadak said.

"Definitely!" Jack elaborated. "The Void is so small that you won't see it, thus the scanners have to be hooked up with the driller very delicately. Also, we don't know how the Void reacts to outside disturbances. We still don't know what made the Void on Venus release the energy surge that killed Ludmilla." Jack's thoughts darkened as he thought about Carmen and wondered how she was doing. She should be back at Earth by now and recovering, he thought, if she had managed to hang in.

Zadak pulled him out of his thoughts. "Nothing happened to your crew when you were poking at the Void, Jack?"

"Correct. I hope we can get some more info on that Venusian accident from headquarters soon. Meanwhile we should proceed very cautiously. Pablo, you will assist Zadak with the drilling, while the rest of us explore the surroundings. Let's split up into two teams."

Jack noticed that Kitahari felt uncomfortable with Chan, so he grouped her with himself, and put Joe with Chan, so as to have a senior researcher together with a junior researcher in each group. While hovering over topographic and geologic maps, they discussed the route each team would be taking. Jack, Zadak, and Chan were most interested in a location about 8 km from the Void, which was dotted with clusters of caves. The discussions finished rather quickly and each member settled onto their sleeping pads to slumber.

## 3.3 Excursions

They all got up with sunrise to the smell of hot coffee. The temperature in the canyon had fallen to incredibly frigid levels during the night and stayed that way well into the morning—until the sun rose over the rim of the canyon. It only very slowly warmed up the arctic-cold air around them. Nevertheless the camp was busy with everyone preparing for the day. Zadak and Pablo performed the final touches and hook-ups on the drilling equipment, with the scanning apparatus mounted on top of it to make sure that they were not drilling too deeply. The scouting team with Chan and Joe left first, in the direction of the caves. Jack was still discussing the scanning procedure for the Void with Zadak. Even though there was a scanner mounted on the drill, he wanted constant surveillance with a nearby handheld scanner to monitor any changes or fluctuations in the Void. After he resolved this matter, he left with Kitahari in the same general direction as the team led by Chan.

Chan and Joe made quick progress toward the cave area despite the rugged terrain. They found several large lava-tube caves. Before entering the cave which had the largest opening, they called in, and Jack gave them an okay to proceed. The lava tube was gigantic, about 15 m in diameter—much larger than lava tubes on Earth.

"These caves are so large because of Mars' lower gravity," Chan marveled as the opening slowly swallowed him. They went in about 50 more meters and found several cavities branching off the main channel.

"Which way?" Joe asked.

"Let's choose the one that appears to go in the deepest with the lowest topographic elevation. I really would like to find some water. It should be there in the topographically deep areas. Let's keep an eye open for moisture."

"Okay."

Chan and Joe advanced deeper into the cave labyrinth.

*****

Meanwhile Jack and Kitahari advanced toward the caves on a slightly different route, so they could check out a nearby salt playa. The ancient salt left a whitish crust that was easily identifiable from space. Jack had noticed it when they were approaching yesterday. When Jack and Kitahari got near the area it was obvious why. The bright whitish color was in stark contrast to the rust-red surrounding Martian soil. Still, the location of the dry salt lake in an outflow channel within Valles Marineris was puzzling. Was upwelling groundwater intersecting the valley floor, evaporating, and leaving the salt crust behind? Or was it a remnant of an ancient surface water body? The team proceeded to the center of the playa. Gypsum crystals protruded from the ground everywhere, and gave the area a mysterious, sparkling appearance. Some formed patterns, reminiscent of skeletons from large organisms like whales. Jack took several pictures of odd formations with his digital recording device as they walked by. When they entered the center of the playa, Jack set down his pack and took out his shovel: he started digging. At about half a meter depth he stopped and turned to Kitahari. "No moisture. Everything here is bone-dry."

Kitahari was curious. "How did the salt crust and gypsum crystals form?"

"I can't be certain. We will take some samples to find out. It looks to me that groundwater evaporated some time long ago, forming the gypsum crystals and the salt crust."

"Similar to White Sands National Monument near the New Mexico Space Harbor?"

"Yes, not so much White Sands, the sand dunes are actually pure quartz, but more like the big salt playa within the National Monument. It's called Lake Lucero, I think."

Kitahari retrieved samples of gypsum crystals and the salt crust from various depths and entered some notes into the little computer attached to her left arm. Then they continued on their way to the cave area. After several more minutes of walking they were back again in the usual rusty red Martian soil, but every once in a while they still noted some white-yellowish salt and sulfate minerals interspersed with the Martian soil.

*****

Chan and Joe made good progress and had advanced several kilometers into the cave complex. Joe's GPS-Galileo location unit indicated that they were 280 m below ground level. Their instruments also indicated a rise in humidity and air density. They followed these increases as indicated by their atmospheric scanners. Finally, they noticed several stalactites and stalagmites ahead of them. It was an impressive find. It meant that substantial amounts of water had migrated downward through the ground for some time before precipitating into these rock formations—enough liquid water so that it had dripped, not just evaporated instantly. They hastened forward for a closer look. Joe was excited. Some of the stalactites were covered by slime, an organic residue of microbial activity. As they entered more deeply into the cave system, they discovered more and more calcareous structures that were coated with the slime.

"That looks disgusting," said Joe who was having second thoughts about this cave exploration and its present cold, slimy, and pitch dark environment.

"Martian life welcomes you!" Chan grinned. "Do not disturb them. These are called "snottites" for a good reason! The bacteria in the slime combine hydrogen sulfide with oxygenated compounds to form sulfuric acid, but at the same time produce a slimy biofilm to produce their own suitable micro-environment. They use it also as protection from their own waste products."

Joe pointed deeper into the cave to some stalactites from which water was dripping down. "This area must have had quite a bit of water for some time."

Chan nodded, and led them further in. Some of the cavities were quite small and they had to squeeze themselves through the narrow passages. In one larger chamber, they saw a hole with a vertical drop of at least several tens of meters. They couldn't make out the bottom.

"I'll check it out," Joe said.

They untied the ropes from their gear, knotted them together, and tied the end rope around a boulder. Then Joe descended slowly. Unfortunately, their rope was not long enough and Joe couldn't reach the bottom. At the bottom of the rope he took out his scanner and measured air humidity levels slightly above fifty percent. Then he climbed up and told Chan about his measurements.

"We need to return," Chan said.

They marked their current location on the map and then started their ascent, backtracking their steps the same way they had been coming in. Joe was very glad he had recorded their path with the GPS-Galileo unit. The caves really were a maze of tunnels. Several times they took wrong turns and were forced to backtrack a bit to get back on the path.

*****

Jack and Kitahari trotted over a boulder-filled plain toward the cave system. Suddenly, Kitahari stopped. Jack turned to her. "What's up?"

"I played a bit around with the element and mineral scanner. Over there, at 2 o'clock, the scanner indicates a collection of metals and some sort of organic material!"

Jack examined the scanner. "Hmm", he said. "That is indeed very odd. I don't know of any mineral assemblage that has this kind of signature. And the organic part doesn't make any sense at all." Perhaps a malfunction, Jack thought as he tested the scanner against his gypsum samples. "But everything seems to work alright."

Jack slapped the instrument several times against his palm—the age-old universal cure for ailing electronics. Nothing changed.

"What should we do?" Kitahari asked. "Proceed to the caves?"

"The caves can wait. After all, the major part of our mission is exploration. And we have some kind of a mystery here. Besides, Chan and Joe are already checking out the caves."

Jack radioed in their finding to the base camp and let Zadak know about their detour. He tried to reach Chan as well but couldn't get through. It didn't worry him though. He figured they were probably exploring a deep cave system and were out of reach.

Then they followed the instrument's metal detection toward the anomaly. The terrain was rough and it required a lot of climbing over fields of rocks and boulders. They saw something reflecting light ahead of them and headed straight toward it. On their way Jack found a metal piece the size of a hand, which was heavily blackened at its edges. Jack and Kitahari examined it from all sides.

"It definitely looks like a piece from a spacecraft," Kitahari said. "Did anything crash around here?"

"Not that I know of, but we should know soon. Perhaps this is one of the earlier Martian probes. Many of the earlier missions failed and the probes crashed on Mars. I bet the reflection in the front of us is the spacecraft. Wouldn't that be an exciting find?"

They proceeded and indeed discovered the jumbled wreckage of a sizeable spacecraft scattered along the side of a rocky bluff and into a sandy ravine. Jack and Kitahari picked through the wreckage looking for markings to identify the craft. They found some Chinese letters etched into the metal and a red star that was barely recognizable. Jack called in to inform the base camp, but Zadak did not know anything about a crashed Chinese spacecraft in the area. Jack remembered that a couple of Chinese missions were launched to Mars about thirty years ago, but that no further Chinese missions had been launched after the American-European human mission landed successfully on the Red Planet. Zadak suggested a call to Morty at the base station. Jack did so and described the wreckage to Morty in detail, while Kitahari was exploring the wider area of the crash. Morty's excited voice vibrated through the communication relay. "Quite a find you made there. It can only be the Mao-I spacecraft."

"I don't recall such a mission."

"It was not very publicized. Remember, the Chinese wanted to beat us to making the first human landing on Mars. The American-European mission took so long because our objective was not only to televise the first human walking on Mars, but to expand the robotic station and establish a permanent human base. The Chinese actually beat us with getting the first human to Mars, but something went wrong, and the mission was lost. There were reports that the Chinese wanted to make a big public event out of it, but they were aware of the high risks, and thus didn't announce anything beforehand. And after the mission failed, they didn't want to talk about it, and of course our governments, for political and diplomatic reason, did not inquire either."

"Wow, that's quite a story. I heard rumors about it, but thought that it was only tabloid junk."

"No, no, that was for real. Say, do you see any human remains?"

"Not that I noticed. But let me look around. "Kitahari, come in!" Jack signaled. "Kitahari, come in!" he repeated.

The radio remained silent. A feeling of dread crept up on him. He scrambled up the bluff and ran to the area where he saw Kitahari last. He climbed to the top of a large boulder and saw her kneeling down a few meters ahead of him, in front of a human body. Jack jumped off the boulder and ran up to her. Kitahari stared at the body, which was encased in an antiquated Chinese astronaut suit. The helmet was cracked and from the inside two narrow eyes stared back.

"Look at him!" Kitahari said. "It looks like he just landed yesterday."

"Yes, absolutely no decay. He must have come down many years ago, crash landed, somehow miraculously made it out of the wreckage, and got to here before he died."

Jack helped Kitahari's stand up. They walked back to the wreckage. Jack called back the station and asked for Mortimer, while Kitahari sat down. "Morty, you were correct! We've found the lost human mission. We found the Chinese astronaut as well. He managed to get out of the crash alive. We found his body nearly a hundred meters from the crash site. He didn't look injured. I figure he just waited for his air to run out and then must have suffocated."

"Wow, Jack! Oh my God, this is incredible! What should we do? Well, I will contact the Chinese officials and ask them how they would like us to proceed. I guess we will have to re-write the history books! The first human walking on Mars was Chinese even if he only survived for a short while. At least in death he will be famous. I'm impressed, Jack. Only a few days on Mars and already you've made a huge discovery."

Jack silently reflected on the portents of his discovery and saw Kitahari looking up at him. "He came all the way here. Knowing it was a suicide mission only to die a few moments after getting here," Kitahari said flatly, but her eyes spoke other troubled thoughts. "Is that bravery?"

Jack didn't know how to answer her. The resolve of the Chinese to accomplish a task at any cost was certainly exceptional. He remembered something that Hauser, his martial arts instructor at the Complex, had said to him: 'You need to be ready to die,' he had said 'otherwise you are not 100% committed'. Jack nodded his head unconsciously at the memory before answering Kitahari. "I don't know what it is, Kitahari, but it's definitely something special. We'll leave things like they are and return to the base camp."

Kitahari and Jack took some time to further examine the crash site, but didn't notice anything more that was unusual before they decided to make their way back to the base camp. Chan and Joe had already arrived. Jack went to talk with Zadak first, to get a report about the progress of the drilling team.

"Contrary to your exciting findings, our job was pretty dull," Zadak informed Jack with a twinge of disappointment. "We drilled 8.5 m into the subsurface. Our models indicate that the Void is located 12.02 m below ground. We plan to excavate to a depth of 11.9 m. That should take at least three days given all our safety precautions. Of course, during all of that time we will constantly scan and record the pulsating pattern of the Void."

"Did you discern any pattern in the pulses?" Jack asked.

"Not really. We ran it through the computer, but nothing so far."

"We still need to discuss tomorrow's plan."

Then Chan broke into the conversation. "I would like to go back to the cave system as soon as possible. My preliminary analysis on the snottite samples through the double mass spectrometer—liquid chromatograph indicates various lithotrophic organisms using inorganic carbon, sulfur, and iron in their metabolic pathways. Plus heterotrophic organisms thriving on organic carbon. I'd like to take some more samples. In addition, we noticed a large rise in humidity in the lower cave system. We couldn't explore further, but conceivably, we may actually have pools of liquid water down there."

Jack agreed, but wanted to complete the drilling first before exploring the cave complex further.

"But what do we do with the crash site?" Zadak asked.

Yes, there was the issue with the Chinese astronaut, Jack thought. He didn't know what the Chinese government would decree and he wanted to have a few people present to accommodate their wishes if necessary. "Nothing before we hear back from headquarters," Jack replied. "All right, let's continue with the drilling and focus on the Void. That's what we are here for."

## 3.4  An Astronaut's Funeral

On the next day they continued the drilling and carefully advanced millimeter by millimeter toward the Void. The progress on drilling and scanning the Void was excruciatingly slow, mostly due to all the safety precautions that headquarters put in place after the Venus accident. Jack and his crew knew that they had only a limited time in which to reveal the secrets of the Martian Void but they were also aware of the risks and didn't want to rush or make a mistake. They recorded the energy signature and pulsating pattern of the Void and started to cross-compare it with known emission patterns.

They learned from Morty that the Chinese government had finally decided on the issue of Jao Ling, their lost astronaut. They asked Jack and his team to bury him in the Martian soil. Jack would have preferred to complete his analysis on the Void before presiding over a burial, but in the interest of political relations and common humanity he would perform the burial tomorrow. It wouldn't look good if he just let the man lie in the dust for a few more days while he dug a pit. Jack remembered the prophetic words of Dahai back on Earth. Now, he had to give his first funeral. But at least, it was not for one of his crew members. He was very fortunate, he thought, that so far they'd had no major problems or accidents. He had also learned, in a communiqué that Morty relayed to him, of a message from Mike Hang stating that Carmen had made it safely home to Earth and was recuperating and undergoing reconstructive surgery. Jack felt relieved deeply, like a great weight had been removed from his soul. He knew that he was concerned for her, but he hadn't realized exactly how stressed he had been about Carmen's ordeal. At any rate he was much more chipper and energized than he had been in a long time.

That evening and for the next few days a dust storm swirled over the base camp. It quelled his funeral plans and put an end to exploration trips of the surrounding area. The crew spent most of the time analyzing the energy pulses from the Void and tried to correlate those to anything they could think of. The most straightforward issue was to determine the absolute strength of the signal. It had 2.748 times the strength of the one from Venus. Zadak and Chan had a vehement argument about what this meant. Zadak argued that this means that Mars had 2.748 times the biomass of Venus. Chan argued that the correlation could not be that easy and one would have to account for different types of life, especially in regard to complexity. Also, he argued that it would mean that one thousand times more life would have to be hidden in the subsurface than had yet been discovered near the surface, in the form of lithotrophic microbes and lichen related organisms. Zadak countered that this could easily be true, as most of the biomass on Earth is microbial and hidden in the subsurface. Jack enjoyed listening to these scientific arguments, even if he chose not to participate. At the very least this was good evidence that the crew didn't have any big interpersonal problems. After a week the storm died down, although visibility was still poor due to a thick dusty haze of micro fine dust. Jack took the opportunity to address the burial. They set the scanner on auto-recording and the entire team hiked to the crash site. The dust storm had not affected the sheltered crash site and Joe and Chan set about digging a shallow grave. No one said much. It was a solemn affair. Zadak and Pablo carried the body to the hole and covered him with the rusty red Martian soil. When they were finished, Kitahari laid a cross of rocks on top of the grave.

"I doubt he was a Christian," Zadak commented.

But no one reacted to his comment and she went ahead anyway. Then they assembled around the grave, bowed their heads, and Jack said:

*We come together today to mourn the loss of one brave human who ventured out to the frontier of science and exploration and gave the ultimate sacrifice, his life, for the discovery of new worlds. His truest memorial will not be in the words we speak, but in the way he led his life and in the way he lost his life—with dedication, honor and an unquenchable desire to explore this mysterious and beautiful planet.*

*Jao Ling's sacrifice confirms that the future of space exploration is not free. It is a story of human progress and struggle against all odds. We have lost one invaluable hero of space exploration, but will carry on as humankind has carried on before on our home planet when horror, terror, and despair bloodied our way. Exploration and discovery is the truest essence of humankind and we will not shy away from challenges that await us.*

*Jao Ling, may your soul find peace in this alien world!*

Jack finished his eulogy and stood motionless and silent for several moments. When he felt the time was right he broke away and they all walked silently back to the base camp. Jack radioed Morty that the funeral was over and that the crash site was free for further investigations by Morty and his crew whenever their schedule would allow. Jack was relieved that this part was over. He wasn't completely comfortable having to lead the ceremony, but that was one of the burdens of command.

## 3.5  Discovery in the Lava Cave

During the week they continued to analyze the pulsating signals, looking for correlations to Martian biology, which required frequent contact with Morty and his crew. Their base camp was in one of the deep canyons of Valles Marineris and their ability to contact Earth directly was poor at best. There was the Mars Orbiting Relay Station that transmitted and received messages to and from Earth, but even satellite reception was sporadic in the bowels of that canyon. Thus, all major information was relayed through Morty and his base station.

The slow progress of their analysis and the feeling of being cut off from civilization lowered their morale. After four weeks of work, the crew grew increasingly frustrated.

"We need to fully excavate the Void and examine it. Jack, you did it on Venus, and nothing happened," Zadak argued.

"What would that help?" countered Chan. "The Void is so small that you can't see it anyhow and we can pretty much outline it with the shallow soil layer above!"

Jack could see the impatience in Pablo's and Kitahari's eyes, but they didn't say anything. Finally, Jack said, "I will make a request to fully excavate the Void. We need to make more progress. We need to improve our understanding. Our time is limited and we have a much more difficult job ahead of us on Titan. If we can't solve this here, how can we be successful on Titan? And while we are waiting on the response from headquarters, we take a break and finally explore the cave system that Joe and Chan found."

The camp became busy again at sunrise. A westerly wind picked up in the early morning, but not sufficient to interfere with the day's plan. Jack, Zadak, Chan and Kitahari left early toward the cave system, re-tracing the route used by Chan and Joe weeks before. They noted the location of the cave entrance on their GPS-Galileo locator. This time, when they entered the cave, they were well prepared, with sufficient rope, lighting, and scanners. They reached the vertical drop that stalled Chan's and Joe's progress nearly a month ago, and secured the rope to a convenient stalagmite. Chan went down first, squeezing himself through small rock openings until he reached a large chamber. The chamber was enormous in size and he couldn't see the adjacent walls. He continued to descend further until he felt some kind of changing resistance beneath his feet. He was apparently standing on something fragile. Chan removed the scanner from his side pocket.

"What did you find?" Jack shouted from above.

"Give me a moment. It's water ice." At just that moment the ice beneath his feet broke, submerging him in the water underneath.

"Are you alright?"

Chan straightened up. The water reached up to his hips.

"Yes, everything is fine. We wanted liquid water—and now we have it! Nearly a meter deep. Follow me!"

Then the rest of the cave team climbed down the rope. They found themselves in a middle of a large chamber. With their flashlights they could barely make out the walls of the chamber.

Kitahari spent a few moments analyzing the cave system with her scanner and then reported on her readings. "The drop is fifty-seven meters in elevation and the chamber has dimensions of about seventy-five meters by two hundred meters."

Then they waded to the side walls that provided a small shelf of hard mud to stand where they got out of the water. They noticed that the water was flowing and decided to follow the flow as far as they could. They progressed along the edge of the cave. The cave narrowed for a while and the party was forced to wade through some deeper water, but then it opened up again into an even larger chamber. There were artichoke and chimney-like structures up to half a meter tall in the extremely clean and transparent water. Nothing organic appeared to be stuck to these structures, but they clearly were too complex to be inorganic. Zadak tried to wade across the width of the pool, but it was too deep: besides, he didn't want to damage any of the structures by accident. Sensor readings indicated the pool chamber was just over 150 m in diameter and the pool reached depths of up to 7 m in places. Kitahari scraped off some material from the structures and put it into a sample collection bag. Closer analysis showed that the structures were porous and brittle, and disintegrated easily upon being touched. The group minimized their intrusions because they didn't want to disturb or damage these delicate structures unnecessarily. Then they marked the location of the chamber on their map for future reference.

Chan asked Kitahari to note down the exact measurements of the chamber.

"Will do," she replied and continued, "This is pretty exciting! How important do you think our discovery is?"

"I don't know. That depends a lot on what these structures are. We will know more after we run some analyses on the samples you collected. But at the very least we have discovered an important source of water for future exploration."

They affixed three large lights at the side walls and filled the whole chamber with a bright light. The average water depth in the lake was 2 m with some of the larger structures nearly touching the surface of the water. The structures had distinct geometric shapes and included cone-shaped seepage structures up to 1 m or more in height with hollow internal conduits that were open at the top of the cones. The other dominant structures were dense artichoke-like calcite "leaves" on deep-water mounds taller than a meter.

Chan and Kitahari were busy taking additional samples, while Zadak and Jack discussed the structures.

"These structures are quite intriguing, aren't they?" Zadak asked. "At first I thought these might be like marine stromatolites; those are the oldest known fossils found on Earth, and they date back some three billion years. They were formed by photosynthesizing cyanobacteria and other microbes, but these things here are clearly quite different structures."

Jack agreed. "Yes, those geometric forms imply functional properties. The hollow internal conduits do not extend through the bottom, which would indicate that they feed from the top. It is puzzling how the microbes can survive in the tubes. That can't be a trap structure, or? This looks a lot like a primitive sponge-like organism."

"I don't think so. We've learned a lot about microbial communication in the last half century. Microbes can swap genes very efficiently, especially if they are related or live in similar environments. In a way these microbes could be so deeply interlinked with each other that it would be difficult to tell whether they are working as zillions of individual cells or as one single, integrated organism."

"But you found a lot of different DNA from these structures. Wouldn't that imply that these microbes, even if they cooperate closely, act as independent species?"

"Assuming conventional biology, I agree," Zadak said. "But we have to keep open minded. Look, even in our body we have so many different cells; blood cells, skin cells, brain cells, organ cells. They are all quite different, but interact intricately with each other. As an emergent property of evolution they, collectively, produce you and I, human beings, but each single cell is alive by itself as well. This does not even take into account the billions of microbes that inhabit our body that we need to survive. We don't completely understand them. Most don't seem to have a purpose, but many of them are beneficial, although some of them are pests. If you removed all of them from your body you would die. This goes beyond symbiosis. Complexity is a challenging issue and our assessment may depend more on the bias of the observer than anything else. I don't say that those microbial structures are on the level of a macro-organism, but it brings up the question of how much complexity and structure can be achieved by a multitude of communicating microbial cells. This may be critical for the evaluation of any newly-discovered alien life and its organization. The long-term intimate co-habitation of these microbes is well suited for intercellular chemical signalling and functional coordination. It may increase their opportunities for genetic exchanges. Their mutual interactions could be encoded into a community-wide genetic program. Maybe this is a different progression toward functionality that evolution has selected for on Mars. It is definitely different from the one with which we are most familiar, namely Earth's own."

Jack thought for a moment. "If we don't understand our own cousins here on Mars, we will have an enormous challenge when we get to Titan. On Titan, life should be really weird, since it is most likely of a separate genesis and not related to Earth life."

"Absolutely!" Zadak agreed. "The alien organisms we know about on Mars and Venus branched off from Earth life about 4 billion years ago, but they still have the same type of DNA and similar cellular machinery. We can only make wild guesses as to what we may encounter on Titan and beyond."

"We need to get some of these samples to Mortimer and his crew. They have more powerful instruments on the base station."

Jack turned to Chan. "Are you aware of any similar structures discovered in Martian caves?"

"Yes, there were some similar structures discovered near the hydrothermal vents of the Elysium rise, but not as large and elaborate. And I believe we have similar structures in some lakes on Earth. Pavilion Lake in British Columbia, Canada, for example. The water temperature at Pavilion is slightly above freezing as well, but the water here is more saline."

"Really? That's intriguing. Let's get some samples for Morty and his crew."

The party continued their observations for a few more hours at the cave lake before it was time for them to work their way back to base camp. En route, Jack contacted Mars Base Station and asked Morty for a noon extraction the following day.

## 3.6  Revelations

The next day Jack and Zadak made their way to the extraction point with the samples and all the data they had collected. Marina was there waiting for them when Jack and Zadak drove the rover onto the mesa. They boarded the hopper quickly and wasted no time on pleasantries. Marina took off and turned toward the Martian Base Station. Once they were in the air Jack chatted a bit with Marina about their discoveries and results, but he really wasn't in the mood for conversation so they soon lapsed into silence.

Jack was eager to find out if NASA headquarters had made a decision on their suggestion to drop some of the safety precautions in their research approach to the Void. He also needed to discuss their strategy for the next few weeks with Morty. With things going as slowly as they had, they needed to think of something else if they were to complete their task here before it was time to leave. They would also drop off the samples collected from the lake for the station crew to analyze. The base station had much more advanced analytical facilities than the base camp. The instruments Jack's crew possessed were assembled to analyze the Void, not microbial samples, so the base station team should most likely be able to find out even more.

Morty had arranged a conference call with Mike Hang, Jack, and himself. Jack arrived a bit late at the base station, and Morty and Mike were already talking via the telecom link. Jack noticed that Mike Hang looked very serious and a bit fatigued.

Jack entered the conversation and was greeted by Mike. "How are you, Jack?" Mike said with a welcoming smile. "I'm sure you're full of Martian dust by now."

"Yes, definitely," Jack agreed. "More than I'd like for sure. I believe Morty forwarded you a message concerning my request for the full excavation of the Void and for permission to by-pass some of the safety precautions. We haven't made much progress here and I was hoping for some leniency to expedite things."

"I reviewed your request and approved it on the condition that Morty agrees with any of your actions that might pose an increased risk to you or anyone else on Mars."

Jack was astounded. That was too easy for dealing with a bureaucrat! They always want to play things overly safe—to the point of being counterproductive. He had expected to have to argue with Mike in order to strike a deal.

The NASA administrator continued. "Morty and I agree that you need to find out as much as you can about the Void. You only have a limited time left; one month at best. Therefore I agree with you that progress is essential and if that means a somewhat higher risk, well, so be it. You need to continue onwards to Titan shortly. But I should advise you that someone else from headquarters might be assigned to guide your mission at that time. Either way, you are trained well and I'm sure you will succeed. Morty can keep you posted on some of the latest details and developing news. I hate to cut this short, but I have to run to a senior staff meeting at the White House. Best of luck, Jack!"

Mike Hang vanished from the monitor. Jack looked at Morty with a big question mark on his face.

"That's not the Mike Hang I know," Jack said. "He's hiding something from us or is deeply troubled by something."

Morty shook his head and swallowed loudly. "It's the Kuiper Belt Object out there. They recorded radio and light emissions from that KBO that's heading toward us. They can't explain it as being either natural or random processes." Jack's heart dropped as he listened to the news and Morty continued. "They are ninety-nine percent sure that those signals are artificial, of intelligent origin! You can imagine what huge political chaos that is causing. The UN is trying to unite the major powers, but it seems that every government has its own ideas of what to do. Your government in the U.S. is leaning toward leaving it up to the military and Mike may be replaced by one of their generals shortly. If that happens, NASA is on its own, of course, and won't have the direct support of the European, Chinese, Russian, and other space agencies. Mike is under extreme pressure to send you right now into deep space to intersect the KBO. But we both agree that this would make no sense. The most prudent decision is to wait and figure out what these Voids are about. There must be a connection. There must be. We need more information on those Voids, Jack, and fast!"

Jack was stunned: alien life—and intelligent! Could that really be true? He recalled his jokes about this while on Earth. When life was found on Mars and Venus people grew confident that they were not alone in the Universe. But then, after no new discoveries were made despite all the new remote sensing equipment and SETI initiatives in the last few decades, people appeared to become resigned to the idea that all life that might be "out there" was probably at the microbial level. After all, we might be the only intelligent life forms, at least for vast stretches of space. Now, that belief was turned upside down. Well, it was not 100% confirmed—there was still some uncertainty. Nevertheless, a truly stunning development, he marveled.

Morty and Jack continued to discuss the recent developments and Zadak joined them after he had dropped off the samples with the base station crew. Jack told Morty about the cave lake that they found and suggested they might want to establish a little outpost in that area. Their discussion was helpful but mostly it just raised more questions and concerns.

When it was time for Zadak and Jack to leave the base station, they migrated to the gas hopper without fuss or fanfare. Marina piloted them back to the mesa rendezvous point and dropped them off. The hike all the way down to the base camp was a quick affair—much easier this time without being burdened with too much equipment. Morty had blessed them with a bulging bag of fresh vegetables from the station greenhouse. Jack was pleased with the gift. It would be good for crew morale and infinitely better that the dried food they had been eating.

When they reached the base camp, Jack informed his crew about the recent developments. They launched long discussions of numerous questions and concerns—for none of which he could provide a satisfactorily answer or a solution. He told them that the most intelligent thing to do was to focus on their work and to find out as much as possible about the Void.

The next days were busy. Everyone tried to focus on the work, but the knowledge of the anomalous and possibly alien object heading toward Earth was a nagging distraction for everyone. With the authorized extra effort the Void was fully excavated, and without encountering any problems. They elevated their testing procedures to more interactive levels. They moved different materials through the Void and measured the effects that the Void had on the materials, and analyzed for any residual imprints left by the atomic replacements. As on Venus, the Void was only a few microns in diameter, their best measurement gave a volume of 2.6 cubic microns. Whatever material they moved through the void, the atoms were replaced in a perfect manner, without impurities. Elements were exchanged in such a way that it left the replaced matter, rock or artificial material, in a completely pure state. This elemental exchange even worked for organic and biological material and Pablo pointed out that a larger Void like the one on Earth with a diameter of about 10 cm might be used as a "miracle treatment" in the healing process for physical injuries. Of course, that was politically impossible since no one would ever be allowed close to the Void. As far as they knew, the U.N., Israelis, Arabs, Americans, Europeans, Russians, and Chinese were still arguing about whether to drill down to the Void underneath the El Kas Fountain. In reality, the Void on Earth seemed to be the key, at least to Jack. And if those at home could finally agree on some procedure for studying it, he wouldn't be stuck here at the end of the world. He shouldn't think negatively, though. Out of all the people in the world, here he was, leading the mission—and it was up to him and his crew to do their part to solve this riddle. And the new, possibly alien, thing heading this way only raised the stakes. It was not pride that made him see it this way for he certainly considered it an honor to be here, not his due. No, it was something else, some elusive realization that he just couldn't wrap his consciousness around.

His musing was interrupted by an intense and loud discussion between Chan and Zadak, who were considering the philosophical implications of the Void. How could this thing, or lack of thing to be more precise, 'know' how to re-assemble anything in a perfect order? Jack and his crew tried to figure out how to distinguish the emplaced atoms from the originals, but they failed to identify any special property. They made no further progress on understanding the pulsating pattern and were unable to link it to any known natural phenomenon. All their attempts to analyze the Void were foundering in descriptive work and time was slipping through their fingers.

Jack and Zadak had long discussions about life and the universe, and how the Voids related to it, which proved challenging but interesting since Jack, as most other scientists, believed in the reality of the physical universe, while Zadak doubted it. "Maybe someone put the Voids here? Maybe, the computer programmer?"

"What?" Jack muttered, puzzled.

"Maybe we are all living in an elaborate computer program! You know, there is actually quite a bit of evidence for this. Consider the observer effect in quantum mechanics. Extrapolate the effect to the macro-scale! It would mean that if a rock falls onto the Moon and no one observed it, it never happened. Or more accurately, it was never defined. How would you better save information and space in a computer program?"

"What are you saying?"

Take another example. Aren't you puzzled about the information entropy effects around black holes? The reason that radiation, Hawking radiation, is emitting from black holes is solely based on information entropy. Jack, information—not physics!—rules the universe. Doesn't that sound like a computer program? And then the anthropogenic principle, the absolute fine-tuning of the universe to make it just right to allow stars and planets to exist, life, us,…

Jack interrupted him. "Yes, what about us? What are we? And the Voids?"

"Us?" Zadak laughed. "We are only elaborate subroutines of a large 3-D computer program. And the Voids could have been put in by the programmer to create the subroutines. Maybe the reason that we don't find a physical basis is because there is none."

"Absence of evidence is not evidence of absence," Jack blurted out. He couldn't accept Zadak's views. But he asked himself whether it was only because he didn't like them personally or whether his scientific beliefs were founded on scientific evidence. Zadak was no doubt a crystal clear scientific thinker and his ideas made some sense.

Meanwhile, Chan led several small groups to further explore the nearby cave systems. They found a few other subsurface lakes similar to the one discovered before, but for the most part much smaller and with little microbial overgrowth. One hope was to gain a better understanding of Martian life and somehow link it to the Void as Jack had done for Venus. Of course, it also helped the crew's morale to explore and assist in mapping resources for a future expanded human colonization of Mars.

Nevertheless, the inescapable fact remained that they had been delving into the mystery of the Martian Void for over six weeks with little results. Jack and his crew grew increasingly frustrated with each passing day. He lamented that they hardly knew more than he knew back when he was on the Venus Orbital Station. It all seemed like a horrible waste of time. They had made some other exciting discoveries, but no real progress on the actual objective, the understanding of the Void. He went over Carmen's report for the tenth time but apparently she did not have any clue at all what Ludmilla was doing to cause the energy burst that killed her. Maybe she wasn't doing anything at all and the Void had just spontaneously destabilized? Jack left the tent and walked over to the scanning apparatus and addressed Zadak. "Any new insights?"

"Nothing. It's extremely frustrating. We must be doing something wrong, or approaching it from the wrong angle. I'm out of ideas, Jack, and I can't think of anything more we can conceivably do here!"

"By now we should have at least identified the full pattern," Jack agreed, mirroring his disgruntlement.

"The only thing I can say for sure is that there is a regular pattern interspersed with random signals. But then we cannot pinpoint the exact regular pattern within the whole frequency spectrum. There seems to be a connection between the regular pattern and the random signals. I know that this does not make any sense. From the part that we can identify as a regular pattern we ran some correlations with known cyclic patterns. The highest correlation we found was with the reproductive cycle of an Australian ant species from 2003 to 2014 overlain on the vibration of a highly pressure-shocked carbon atom in a strong electromagnetic field. This is only nonsense, of course. I guess your Venusian Void was easier to figure out."

"Based on the Venus analogy, I would expect that it should be related to the life cycle of Martian life. But we don't have any idea or indication of what that cycle might be. The Martian life cycle might repeat over a span of centuries or thousands of years. If that is the case we might never see it."

"That could very well be. We tried it, but we know so little about life on Mars. Only the tip of the iceberg. The majority of biomass on Mars is in the subsurface and we don't have much clue about it at all."

"What do you suggest then?"

"Nothing I can think of. We went through the data bases of Martian life at least five times, forward and backward. We don't have the resources to make any major headway in the short time we have here. I think we should pack up now and head to Titan. I don't see a realistic chance to do more here. And Morty and his crew can continue to work on this Void in our absence."

Jack was silent for a while and then nodded his head slowly. "I think you're right."

## 3.7 The Arivees

In the evening Jack called Morty and told him about their status and the suggestion of suspending their research here in favor of spending more time on Titan to find life and study its Void. Jack hoped that life on Titan would be close to the surface and thus possibly easier to discover, study, and relate back to the emission patterns of the Void. Morty sat back for a moment and assured Jack that he would forward his suggestion to Mike Hang and talk with him about it. That was all Jack could do for now. He would have to wait for Mike's reply.

Jack was just getting up early the next morning when Morty called back. "I just got a hold of Mike Hang and he agrees with your suggestion about heading for Titan. But he also says you can't leave for another week."

"Why not?" Jack asked, wondering about the delay. "We can have everything packed up within two to three days."

"Mike has sent you two additional members to complete your crew for the Titan mission. They are already on their way and are scheduled to arrive in six days."

"What additional crew members? He didn't mention anything to me. Why wouldn't he disclose that information?" Jack was upset and his mind was roiling with the possibilities. Mike Hang thinks that we can't do the job by ourselves, he thought. He's not satisfied with our progress so he's trumping my mission authority. Jack tried to dismiss this line of thinking. Fear and insecurity about his performance wouldn't accomplish anything.

"I'm sorry. That's all I know, Jack. Those are Mike's orders. There is a lot of confusion lately and they don't keep us well-informed anymore. I think their right hand does not know what their left hand is doing, and they are all too busy with their stupid power struggles."

Jack nodded his head but he wasn't satisfied with Morty's response. "I'll report in later," he said coldly before signing off. He didn't like this change of events. You don't just disrupt a mission by adding two new members without consulting the mission commander. It undermined his authority and made him seem powerless.

Jack told his crew to continue working on the Void for the next three days in the faint hope of a scientific breakthrough, but held back the information about the two new crew members. Afterwards they would start packing and prepping for their departure to Titan. He also relayed this information to Garibaldi and Lana so they could have the ship ready for the trip. A day went by, and the next, and another, but no significant insights or breakthroughs showed themselves. The mission that had started out with so much excitement, with the discovery of the underground water and the Chinese wreckage, felt like a failure now. With heavy hearts, they started packing up their gear. They would leave the driller and some of the other equipment in place for Morty's crew to continue the scanning and to perform any analysis Morty felt compelled to do.

The next morning they finished packing up their gear, brought down the tent and made their way back to the mesa. Marina was there, waiting to fly them back to the Base Station. Garibaldi already waited for them when they arrived at the Martian Base Station.

"Everything is in top shape, Sir. We are ready for launch," he excitedly proclaimed. "I'm looking forward to our next endeavor."

Jack was glad that at least one of his crew members was still excited about the mission. He called his senior staff, Zadak, Chan, and Garibaldi, together in his room on the base station and told them that two more crew members would accompany them to Titan and that they would be arriving tomorrow.

"They didn't tell you who?" Zadak exclaimed in astonishment.

"No, not a word." Jack replied simply, trying not to show his emotions.

"What if they are military guys?" Zadak asked, red in the face.

"You've seen too many science fiction movies!" Garibaldi commented.

"And I'm sure the bureaucrats have seen those same movies. Don't you think it seems really odd that they are suddenly sending two additional members?"

"Odd, yes -, but it can mean anything." Chan commented. "Maybe they think we could use some more expertise. We weren't exactly victorious in our endeavors here."

"We will see who it is and then we can deal with it," Jack said.

"Yes, but then it may be too late. You are not forceful enough with head-quarters, Jack," Zadak concluded.

"Don't tell me how to do my job, Zadak!" Jack responded menacingly.

Zadak put up his hands and backed off. "Forgive me, Jack. I'm just concerned," he responded sincerely. "But consider this, I'm not saying we should be rebels, but out here on the frontier we're all alone and we represent humanity. This is a voyage of discovery and science, not a military reconnaissance mission. If there is indeed intelligent alien life heading our way and we're the ones fated to make first contact, then we can't allow a military presence onboard our vessel. If these two new crew members are soldiers or if head-quarters is resistant to give us their profiles then I say we leave before they get here."

"You're talking crazy, Zadak!" Garibaldi interjected.

Jack knew it was time to put a stop to this line of talk. "That will be enough on this subject. Jumping to conclusions is a dangerous game we don't want to play. We will follow orders and wait until our two new members get here. Understood?"

Zadak, Chan, and Garibaldi agreed.

They spent the following morning at the Base Station working on their flight plans and looking over some of their Mars work. Morty called Jack to let him know that the ship with their additional crew members had already established a stable orbit and that the new people would be shuttling down to the station in a few hours. Jack was working out a Jovian bypass trajectory with Chan and Zadak when Morty called again to tell him that his two new crew members had arrived and were waiting in the conference room. Jack excused himself from the discussion and went to greet the two.

Jack stepped into the conference room and stopped in shock: the two arrivals were Carmen and Vladimir. It took a moment for the shock to subside before he could mumble something to the effect that this was the greatest surprise of his life.

"What!?" Vladimir cried. "You didn't get the letter I sent you. Damn Martian postman!" He cursed with his Slavic accent and then rushed forward to embrace his old partner. "Mike Hang thought you could use any help you can get," Vladimir said laughing.

"You bet!" Jack smiled back. Then Carmen came forward and his smile dropped as he scanned her slowly, drinking in every detail and feature of her appearance. The signs of her injury and disfigurement were still painfully obvious.

"How are you doing, Carmen?" Jack said as he grasped her shoulders in welcoming concern.

She smiled defiantly. "As you can see, the left side of my face is not quite how it was and the scars still have to heal, but otherwise I feel up to any task and am happy to be part of your crew."

Jack felt the anxiety he had felt over the last few days drain out of him and reveled in the rush of warmth and happiness. What the hell—if he had to travel to the end of the Solar System and possibly face death and danger along the way, well, at least he was together with his friends and the ones he cared most for. "Let's go and meet the rest of the crew," he said excitedly. He took Carmen on his right arm and Vladimir on his left and headed out of the room.

The crew seemed to be open to the additions, and relieved that they were not military, as the rumors had started spreading. Jack said goodbye to Morty and thanked him for all of his help. During the last one and a half months they had developed a close friendship. And Jack was thankful for it–he felt he needed all the friends he could get these days. Both crews enjoyed a scrumptious feast the night before leaving. Jack knew that with so many mouths to feed they had drained the Mars station of its produce for at least a few weeks, but he wasn't going to complain. It would certainly be the best meal his crew would have for a long time.

The crew of the Mars Base Station said their farewells as Jack and his crew boarded the shuttle for their return to *Deep Explorer*. As they lifted off Jack reflected on the recent additions to the mission. The crew seemed to be welcoming to the additional members and to fully incorporate Vladimir and Carmen into the team should not be a problem.

When they arrived back on *Deep Explorer*, Jack gave Carmen and Vladimir a tour of the ship while everyone else prepped the ship for launch.

"This is what you have to become familiar with in the next few weeks," Jack said.

"We will," Vladimir said confidently.

"I'm curious," Jack got serious and changed the subject. "Do you know Mike's rationale when he selected you to join the crew? He seemed to be so insistent with me that every crew member needs to have extensive training in the Complex. And why wouldn't he tell me your identities—or even that you were coming?"

"I think he had to take more risks." Carmen said. "There is a lot of political chaos on Earth right now and there was no more time for additional training. I guess, in my case, my experience on Venus and staying psychologically stable during the whole time made him feel comfortable enough to send me. They conducted all kinds of tests on me, though."

"In my case," Vladimir took the initiative, "I'm the international add-on associated with the Russian Space Agency, and since you and I had a previous and successful collaboration on Venus, I was a natural choice. I guess he wanted to have an expert team with the best mission experience together. But I believe, too, that Mike was under pressure to act fast. I think there are a number of antagonistic factions in many governments that don't want us out here. These factions seem willing and able to sabotage our efforts. Mike wanted to get us off Earth as quickly and with as little publicity as possible."

Jack didn't like that news. Earth seemed like a chaotic den of disorganization and upheaval. Why couldn't people work out their differences in a civilized manner? But then, when in all of human history had they ever done so?

The launch was set for 0600 next morning. Jack had scheduled a telecon with Mike Hang. In some way he was surprised that Mike was still hanging on as NASA administrator given all the political turbulence that Morty, and now Carmen and Vladimir, told him about. Either way he was glad of it.

"We are ready for launch," Jack confirmed.

"Very good. I wish you all the luck you can muster, Jack."

"We will need it, I'm sure. You helped us out a lot by sending Carmen and Vladimir as additional crew members. Thank you very much."

"I thought you would like that," Mike smiled. "I hope you can accommodate everyone with quarters, since the crew is a bit larger than the ship was originally designed for. Just one more comment. I don't have to point out how important your mission is to all of us here on Earth. We know that we can count on you. You are a good man and you received excellent training at the Complex. Don't forget to trust your conscience, your spirituality, in dealing with whatever awaits you out there. And you will need to make your decisions pretty much independently from headquarters. Just keep that in mind."

It seemed to Jack that Mike was trying to deliver a warning, but just couldn't mention it in the clear. He knew that this transmission was being recorded back on Earth and once again Jack had the feeling that there was something Mike wasn't telling him. Jack wasn't sure how to respond but he didn't want to create a prolonged silence.

"I will," was all he said.

"Over and out," Mike said and Jack's screen went dark. "I don't want to be in his shoes," Jack thought, "…but then I don't know whether I want to be in my own shoes either." Things were certainly getting strange. Jack didn't even want to think about the current politics on Earth. Every government pulling in different directions. Some factions charging that the alien Kuiper Belt Object was nothing more than a made-up conspiracy. Others were proclaiming the end of the world. And others argued to simply blow up the Voids and the KBO with it. It appeared to him that this was all much too chaotic and unpredictable to let him completely trust Earth's leadership. It was a high-risk gamble as to which Earth faction and philosophy would win. Maybe that was what Mike meant by trusting his own judgment; after all, Jack and his crew would be the only humans close enough for an inside look. He tried not to let such dark thoughts weigh on him.

On an impulse, Jack called Master Lao and was surprised how quickly Lao appeared on the telecom link. "Just wanted to say good-bye since we will be leaving for Titan in a few hours," Jack said.

"I'm sure you will have a great adventure! And you are well prepared, too."

"I'm not so sure about this," Jack confided. "It doesn't seem like the Mars expedition was very successful. We really don't know anything more about the Void than when we got here. And I don't think headquarters is very happy with our results."

"Oh, you still haven't fully grasped the most important thing!" Master Lao chided. "Whether you can determine the size of the Void to a decimal point more or less, does not matter. And what headquarters thinks matters even less. It doesn't even matter what I think. What is, is. That's all. Don't wait for any feedback, positive or otherwise. Inside yourself, you will know what the right decision is, and that's what you have to do! And even if the whole world is cursing you for that. It does not matter."

Jack tried to absorb Master Lao's wisdom. "At times it is so difficult to decide what approach to use. Yes, I know deep inside, the Void is something more than a technical phenomenon, but how do I approach it? What can I do to find out more?"

"I don't know the answers to these questions. The only thing that appears clear to me is that your previous approaches did not work out. Perhaps a deeper spiritual approach would work?"

"You mean I should try to communicate with it?" Jack asked dubiously.

"Again, I don't have any ready-made answers for you. The process is the goal, not the result. It is your task to figure it out, put things together, and solve the riddle. You know there must be a reason why the Void on Earth is located beneath Jerusalem. It is a spiritual center, the birthplace of three major religions. If it were impossible to solve the riddle, I don't think you would be in the position you are. But it surely isn't easy. Who knows, maybe you're going after it the right way, but it is not the right situation and proper timing. On the other hand, would it be so outlandish to assume that the Void is something much deeper; perhaps intelligent, something with which you can communicate?" Lao continued. "These are new times. They require radical new thinking. We know there are alien life forms in our Solar System that are associated with these Voids in some way, and now there certainly SEEMS to be an intelligent species flying toward us, also connected with a Void. You are the scientist Jack, not I. Connect the dots."

"I'm not sure I can."

"I believe you can. I'm sure you will do whatever you can. But, of course, we don't know the outcome. Whatever the path is meant to be, it will happen. Have faith in that."

Jack said his goodbyes and signed off. He had not received the solutions from Lao that he had hoped for. And he surely was not prepared to accept just any result and be content with what he had learned in the process. It would be no easy task to sort this all out, he thought by himself. But he agreed with Lao on one thing; in principle, and in fact, he was on his own.

Then he went to his room and meditated in the sitting position that Lao had taught him. Afterwards he went to brief his crew members on the latest mission-related updates that had been sent from headquarters. All of them were already assembled inside the meeting room when he arrived. "It will take two and a half months to arrive at the Saturnian system," Garibaldi said in their first briefing on the journey. "We would be much quicker if we didn't require so much time for accelerating and decelerating. In about a month we'll pass the Jovian satellites but we will keep our distance from Jupiter so as not to expose the ship to too much radiation. Since we'll be near our top speed at that time we will not have any extra time for scanning any Jovian moons or doing scientific explorations."

"Europa was once considered to be a prime astrobiological target, for life underneath the ice cover," Zadak commented.

"Yes, but there is no Void on Europa or any of the other satellites, so no life," Garibaldi replied.

"I believe so, but it still needs to be proven that the Voids are needed for life to occur on a planet." Zadak argued. "They may not always be associated with life on a planet. This is just our leading hypothesis right now. If we can find life on Europa, we could counter that hypothesis. I think we should stop or at least slow down to take more time to look at it. We are a bit ahead of schedule anyhow."

Jack jumped into the discussion. "That is out of the question. We need to get to Titan as quickly as possible. Those are our direct orders from headquarters. Any delay in the Jovian system would not be excused. Zadak, just fine-tune your scanners during the flyby and record whatever you can."

"You know that there are still scientific discussions that there may be a very small and weak Void on Io," Zadak added.

"Come on," Vladimir interjected, "on Io, there is no water. No real atmosphere."

"True," Zadak conceded, "but there are liquid sulfur compounds, lots of volcanic activity, and many lava tubes in Io's crust where liquid sulfur compounds could be available as life solvents. Besides, during early Solar System history Io had a lot of water."

"Well, then, target your scanners towards Io," Jack said. "You are the head science officer. It is your call, but we are not slowing down."

Pablo raised his hand and asked "When do we continue with the training sessions that we had on the way to Mars?"

"Right away!" Jack responded. "When we arrive at Titan I want everyone to be familiar with every system on the ship and the shuttles. We'll put the emphasis on training Vladimir and Carmen, since they are new to the ship. Carmen and Vladimir, I would also like for you to look over our results from the Void on Mars."

The two gave their affirmatives.

After a brief silence Jack added, "I think everyone knows how important our work at Titan is going to be. I want everyone in top shape, physically and mentally, and as best prepared as possible, by our arrival."

Jack looked at Chan who seemed to want to say something.

"With your approval Jack, everyone will have a one hour session in the treadmill every day and I want to see each one of you in a private session with me at least once a week."

Jack nodded his agreement before he continued. "I want any problems, however miniscule they appear, to be reported right away. Any questions?"

Joe raised his hand.

"Yes, Joe."

"I have had some difficulties pooping these last few days. Do I have to report that as well?"

All laughed, including Jack. Only Chan remained somewhat serious.

"All right then," Jack continued. "Get ready for launch." Jack went to his quarters, meditated for half an hour and then went to find Garibaldi.

Jack nodded to him and Garibaldi initiated the countdown for ignition of the primary propulsion system.

At 0600 sharp the engines ignited and *Deep Explorer* accelerated toward Titan.

# 4  Titan

## 4.1  Space 101

The trip to Titan would be long and trying, but the crew wouldn't be idle. Jack developed a rigorous and merciless training regimen that would have every member of the crew knowing every aspect and subsystem of the ship backwards and forwards. Each day would also be accompanied by a period of intense exercise, at least two hours.

On the second day out, Jack encountered Carmen when she was alone in the exercise room.

"You know," he started, "I was thinking a lot about you after we met in the Venus Orbital Station. I know it may not seem very obvious, or even apparent, considering my lack of correspondence. It seems strange but I think that somehow, thinking of you made me finish the training at the Complex. It gave me strength."

"Why didn't you contact me?" Carmen said passively, with little hint of emotion.

Jack suddenly felt uncomfortable with the topic, even though he was the one that had brought it up. He couldn't think of a single response to her question that seemed adequate. Everything he could think of made him feel like a gutless worm and he silently cursed himself for not planning out this conversation ahead of time. "I had to 'fight my own demons' as Lao always said. I don't want you to mistake my silence for apathy or a lack of interest. That is certainly not the case. Honestly. The truth is that I was afraid that any contact would decay into meaningless chit-chat. And I didn't want that to happen. Did you get the note that I sent you after the accident?"

"Yes, it was very beautiful and it helped me pull through. It gave me strength."

"Maybe there can be more between us than just friendship?" Jack said. He moved towards her.

"Perhaps," she responded, but didn't hesitate in her workout. "But not here. I hardly know you, Jack, and more importantly you are my boss and I'm one of several crew members on a long and difficult voyage. That would be inappropriate and make me feel really awkward."

Jack didn't like the response, but he didn't know actually what the hell he had expected her to say given the situation. And of course, she was right. He cursed himself again for bringing up the subject. Rejection hadn't even occurred to him and now he felt uncomfortable, and he could only imagine how uncomfortable Carmen felt. And he had to bring this up on the second day of a long mission where they would be in close contact requiring professional interaction. He felt like an idiot!

Carmen just continued exercising and didn't even look at him. The silence was getting acutely uncomfortable before she spoke again. "Space can play tricks on people's minds, Jack. Especially between males and females. It's space psychology 101. You know that."

"Yes I do." Jack agreed, nodding. He wordlessly backed away and left the room. He felt disappointed, but then, what did he expect? That she would come flying to him, telling him that she always wanted to be together with him? A romantic relationship on a spaceship very possibly enroute to doom?

No, she was right. He had to walk the line. They had to walk the line. Perhaps, at some later time there would be a chance. But not now. The mission must come first. He would have to watch out now, though. He didn't want her to feel awkward in his presence. He made a silent vow that he would treat her as professionally as possible and not bring up his feelings again.

The next few days were awkward anyway—at least for Jack. Whenever he met Carmen he tried to be nice, but also natural and professional. He couldn't tell if she felt as awkward as he did, but she reciprocated in an equally professional and pleasant manner. He took the same approach with the other crew members to show them that their commander was up-beat, motivated, and a good guy. Nevertheless, the long flight to Titan was a challenge for everyone. Not in a navigational or physical sense, but mentally. It wasn't the endless days of being cooped up in a small ship with several others. It wasn't the monotonous dry rations. And it wasn't the boredom. It was the thought that each day took them further from Earth and home and inevitably closer to a mysterious and possibly alien unknown. That fear of the unknown nibbled at everyone's sanity until their nerves were ragged and sharp.

But the crew consisted of professionals. They were all aware of the source of their anxieties and demonstrated exemplary devotion to the mission. The endless training sessions helped combat their fears.

Pablo and Joe expanded the make-shift conference room in the engineering compartment into a comfortable room. They even hung some posters they had smuggled on board, to make a more tranquil atmosphere. And Kitahari had planted four tomato plants, using supplies and seeds she had borrowed from Mars Base Station, to provide a daily dose of 'green love' for everyone. They weren't growing exceptionally well, but everyone admitted that it was nice to have them and it wasn't long before all the little plants had names. Jack didn't do anything to discourage them either, even though having the plants on board was strictly against NASA's ship environmental protocols.

After four weeks they passed the Jovian system. Zadak decided to put the main emphasis on studying Europa, one of Jupiter's icy satellites. There wasn't much time, only a few minutes of good radar screening, since their vessel passed by at nearly their top speed of 0.1 % of light velocity. The surface of Europa looked like a fractured egg shell. Zadak, assisted by Joe and Kitahari, tried to probe with the radar through the ice shell, while Vladimir scanned for the presence of a Void, however weak. The results were not very encouraging. Vladimir reported that there was no sign of a Void—at least not one with a pulsating strength of at least one millionth of that on Mars.

Zadak reported that the ice cover was at least 10 km thick in almost all areas that they probed. Only in one area did they detect higher heat flow and a substantially thinner crust. Kitahari analyzed the impurities in the ice on

Europa's surface and determined that they mainly consisted of sulfuric acid, magnesium sulfate, calcium sulfate, and some other salts. She admitted that she was still struggling to distinguish which of the compounds were from Europa's inner ocean and which were derived by violent volcanic eruptions from the moon Io. There were no indications of organics, but then the surface had been drastically altered by the harsh radiation environment from Jupiter.

All in all they didn't learn much more than they already had known from previous unmanned missions. Zadak was obviously disappointed by the relatively small science return from the fly-by of the Jovian system, but the encounter was still the highlight for the crew on the otherwise monotonous journey to Saturn.

When they were about half an astronomical unit away from Titan, Garibaldi engaged the thrusters to slow down *Deep Explorer*. It would be another two weeks before they would enter orbit around Titan, and the crew continued their normal training routine. Only Vladimir started spending more and more time on the imager to observe Titan and the alien rock heading their way. Every day, Saturn and its moon increased in size on the cockpit screen. Saturn was a glorious site to behold and everyone spent time gazing at its rings and many moons. Titan appeared as a big ball in the distance and detailed cloud features were discernable.

When they were only thirty-six hours from entering orbit around Titan, Zadak used the large antenna to search for the location of the Void, but the dense atmosphere and thick clouds made the determination of an exact location difficult. Nevertheless, after a few hours' work he was able to pinpoint the location to an accuracy of a few hundred meters. It was located near the south pole at the eastern edge of a dark smooth terrain that resembled an ancient lake. He was elated by his success and briefed the group before they entered orbit. Jack was very pleased.

During the briefing Vladimir asked, "Can you see how deep the Void is below the surface?"

"It's difficult to say at this point," Zadak replied. "It appears shallow, very shallow; possibly even on the surface. There is not much attenuation from the overlying material."

"I assume we can improve the resolution while in orbit?" Jack inquired.

"We can possibly narrow it down to within a few tens of meters, but the exact location we can only determine when we deploy our sensors below the methane clouds."

"All right, then we use the remaining time before orbital insertion to monitor the alien KBO and then we will re-focus on Titan when we are in orbit. Our friends on Earth are getting very nervous about the KBO, anyway. What's the status on that alien rock, Vladimir?"

"It is 241 million kilometers away, Jack, and it will be reaching the Saturnian system in five weeks and two days at its current velocity."

"Any new information?"

"We can definitely pick up signals of artificial origin, perhaps communication signals. I'm pretty sure we will have our first encounter with an alien intelligence in less than six weeks. The rock's composition is not that of a typical KBO. It might have come from outside our Solar System. Also, we got the first optical surface images but the resolution is pretty poor. There appear to be various features on its surface, but nothing like on Earth. It looks more like giant access tubes to the subsurface. Also, there are a quite a few thingies orbiting around the planetoid."

"Little moons?" Joe asked.

"No." Vladimir responded. "They appear to be self-propelled without "natural" orbits. I haven't got one in focus yet, but it looks like they are about ten times the size of *Deep Explorer*."

Kitahari went pale in her face and asked. "How many are there? Do they appear to be a threat?"

"This is the same question headquarters had asked me, when I sent my report. There are at least ten of them, and perhaps many more that are smaller or hidden behind the planetoid. Whether they are a threat or not, there is no way to tell. I guess we will find out soon. But if they are hostile…" Vladimir paused to think of his next words. "Well let's just say that the tactical situation would grossly disfavor us in the event of any confrontation."

"Shouldn't we focus on those alien ships rather than exploring Titan?" Chan suggested. "I mean, to me, Titan seems a pretty trivial concern in comparison."

Jack felt several of his crew members staring at him and he rubbed his forehead unconsciously as he thought about a proper response.

"I don't think so," he finally replied. "Vladimir and I will continue observing the KBO or whatever it is, and we will work together with headquarters on a plan for any encounter. Whatever events play out in this mission we will be prepared for them. But there is nothing we can do otherwise. Till the day of the actual encounter, I need every one of you to focus on your tasks here and now. For now, our priority is science. That means finding the Void on Titan and looking for life on Titan. For some reason, the Voids seem to be the key. Everything inside me tells me that. And the more understanding we gain about them, the better we are prepared for whatever will come. Keep in mind that the alien rock has a Void, too, and we will work on it to compare and correlate the alien Void to the Titan Void and the Voids from Earth, Mars, and Venus."

Zadak nodded. "The worst would be if that thing distracted us from our mission."

The crew was silent for a few moments as they absorbed the information and the possibilities of the almost-inevitable future encounter. "Okay then, let's get back to work!" Jack concluded, and the group disassembled.

Jack went to the front of the ship and Lana greeted him as he entered the cockpit. He took a seat beside her and stared into space for a while. "How long until we enter orbit?" he eventually asked.

"One hour, twelve minutes," she responded after verifying their position with the auto-navigation program.

Then Jack went to the communications room alone. He turned on the telecom and had to wait for telemetry to verify transmission alignment with Earth. "Headquarters, please come in. This is *Deep Explorer*."

It took a few minutes to get through, which seemed strange. Jack double checked the solar monitoring array to see if a solar flare was interfering with transmission. No, it was all clear. Then the screen lit up and instead of Fred showing up, or anyone else familiar to him, a balding, about sixty year old man in an air force uniform appeared and responded

"Colonel Haggerty here," it screeched through the static, his voice distorted and crackly.

"*Deep Explorer*, here, Jack Kenton speaking."

"Good morning, Dr. Kenton," the colonel responded, "Hang on for a minute. We are still working on improving the connection. General Bicker will be here in a moment."

Jack had to wait in silence for a few minutes before a skinny, tall man in a green uniform appeared. He had gray hair combed to the side, blue stinging eyes, and a wide scar on his left cheek. From what Jack could pick up through the telecom link, he had his army friends and various NASA personnel buzzing around him like a queen bee. Finally, he moved toward the telecom link and Jack saw him up close. Jack immediately felt an unpleasant aura surrounding this person.

"Captain Kenton, a pleasure to meet you." General Bicker said. His words were pleasant but his eyes were cold and menacing. He quickly continued before Jack could say anything. "Surely, you must ask yourself why you have the pleasure of seeing me on the screen."

"The thought has occurred," Jack responded, somewhat mumbling. He was unsure of the portents, but he quelled the rush of fear and suspicion that threatened to engulf him.

"The president has decided that this is a time of a national emergency and has placed me in charge of NASA. Dr. Hang will remain as one of my advisors for the interim. You will report to me or to Colonel Haggerty, if I'm not available. We will now require daily reports from you about your progress in monitoring the KBO and the Void on it."

"I understand." Jack responded simply but a wash of emotions and concerns swirled in his mind, creating a hundred questions he knew this man didn't want him to ask. For now it would be best if he seems as cooperative as possible.

"What is your status?"

"We are entering orbit around Titan in less than an hour. Vladimir, umm…, Dr. Kulik, will transmit his latest remote sensing results of the KBO object as soon as we enter orbit."

"Very good. We need all the data we can get and have to realize that this KBO may very well pose the most serious and imminent threat to Earth since the extinction of the dinosaurs. We will need absolute and unswerving cooperation from you and your crew. You will receive your orders directly from me in dealing with the threat, and I will receive them directly from the president and the presidential council. Is that understood?"

"Yes, Sir, of course," Jack said simply, with his face blank.

The General made a brief pause as he scrutinized Jack through the monitor. Jack didn't say anything and the General finally dismissed him. "Over and out, *Deep Explorer.*"

Jack sat back for a moment and tried to let the last few minutes sink in. Carmen entered the room, rested her hand on his shoulder. Jack didn't feel comfortable with any physical contact between them no matter how innocent so he twisted away from her as he stood up.

"You look concerned," she said, obviously concerned by his unease. "What's up? Unpleasant news from Earth?"

"You could say that," he responded and told her about the telecon.

"Call the Complex," she suggested. "That is one of the few other direct communications we can get and I'm sure they will treat any inquiry confidentially."

"You're probably right," Jack agreed, "but I wouldn't be surprised if the military was monitoring all of our transmissions. We would have to carefully govern our inquiries."

"You're starting to sound like Zadak," she smiled, making fun of him.

Jack snorted a response trying to downplay his anxieties. "Maybe I'm beginning to buy into Zadak's paranoia." But Jack followed Carmen's suggestion and called the Complex.

Dahai appeared on the intercom and Jack was glad the transmission was much clearer.

"Hello, old friend!" Dahai greeted Jack.

"Nice to see you again," Jack said. "The reason I'm calling is to find out what the heck is going on at home."

"Well, I thought Mike prepared you for this. Since we got confirmation that the KBO most likely hosts intelligent aliens, a lot of people have became very anxious and many countries are now governed by emergency councils, or at least assisted by one, like our country, because of the presumed incapability of political institutions to deal with the situation. What did you expect?"

"I don't think the military will do any better. And we don't even know for sure that the aliens are hostile."

"I agree. Worse still, instead of being united, the world leaders are only scrabbling with each other about the best course of action. Most governments seem set to blow the incoming rock into pieces if they can, while some European countries are tending toward appeasement and possible communication. It's quite chaotic."

"What about Mike Hang, and you, and the Complex?"

"We are all out of the loop right now. You are our best hope for a voice of reason. They don't have any other choice than to listen to you, because you are the only one close to the KBO and you will initiate the first contact."

Jack breathed out heavily. "So everything is up to me?"

"You got it," Dahai said with a mournful smile on his face. "But I find it encouraging that you are out there—you're not an overzealous and fearful… person."

Jack remained silent for a moment or two and then asked "Is Master Lao around?"

"No, I'm sorry, he is in the hospital because of an aneurism, but he should be fine in a few days. I can give him your best wishes."

"Yes, please do so and my best to all the others as well."

"You, too." Dahai concluded the transmission.

Jack gave Carmen an uneasy look and found his own fear and uncertainty mirrored in her face and eyes. She came forward then and gave him an intense hug. Jack didn't shy away this time. He just reveled in the sensation of feeling her body close to his. Then he stepped back and they stood silently for a moment just looking at each other.

"I'll need to let the others know," he said "but not now. They need their minds focused on the mission." He left the room and returned to the cockpit to monitor their approach to Titan, while Carmen went to the launch preparation room.

## 4.2  Landing on Titan

A few moments later the thrusters engaged and smoothly inserted *Deep Explorer* into a polar orbit around Titan. Jack looked at Titan from the cockpit.

Shades of gray were intermingling in front of them detailing the exotic cloud features on Titan. Lana steered the ship into a stable orbit and disengaged the thrusters. After a series of system checks she affirmed their position and ship wide security. Jack activated the intercom.

"Zadak, are you ready to deploy?"

"Ready! All systems are ready to go, the crew is ready."

"Go ahead," Jack said calmly.

The shuttle decoupled from the belly of *Deep Explorer* and veered off toward the south polar surface. Zadak's crew included Carmen, Joe, Kitahari, and Pablo. They were eager to land on Titan's surface after two and a half months of confinement. Jack had decided to have a large portion of his crew exploring the surface and scanning the Void. It would do them good. There should be enough time for the investigation of the Void and for finding any life on Titan, if it in fact existed. But Jack also knew, even before the last communication with HQ, that time was limited and if headquarters got even more nervous, there would be no way to know how long they really had for any exploration. He was not happy that Carmen was going, but there was no logical reason to keep her back, just based on his personal feelings. After all, she was a valuable scientist, possibly with the best expertise on the Voids and life detection. He needed to separate his personal feelings from his professional decisions, and be detached. Jack smiled when he remembered Lao's words.

The shuttle dove into the cloud deck. Jack's breath stopped for a moment as they went out of sight. "Follow their descent with the radar," he told Lana.

"Everything is going according to plan," she responded.

The shuttle shook heavily in the clouds of Titan and the crew could make out a yellowish surface when they were able to peek through the clouds. As they descended, the color of the surface changed to a bright orange. Finally, they could make out various landscapes features. They noticed several dry river beds that were steeply cut in the otherwise softer terrain. Slopes of thirty degrees or higher suggested that the liquid methane or liquid ammonia-water mixtures that flowed in those beds were pretty powerful in eroding the surrounding material. Some sites within the river beds were shiny and implied that the eroding liquid removed any accumulated darker hydrocarbon deposits, leaving only sparkling bluish-white ammonia-water ice. One of the river beds entered into a larger bowl-shaped depression, which must have served as a reservoir for the incoming liquids at some time. The crew observed the landscape with fascination. It was reminiscent of an arctic landscape on Earth exposed to a regional oil spill that happened many years ago and left organic residue everywhere. Finally, they reached their landing location. Pablo extended the landing gear and they landed in a dark-orange colored, smooth area on the edge of a former lake. The landing gear sank a few centimeters into the soft ground surface.

Carmen checked the scanners. "It appears we are less than two hundred meters from the Void." Carmen felt nervous to be so close to a Void again, remembering her last experience on Venus.

Zadak was genuinely pleased. "Everyone, suit up!" he said cheerfully. "It's a bit nippy out there. Temperature readings say it's –180°C. Fortunately there's no wind chill."

Zadak called Jack. "All systems are go. We landed safely and will now exit the shuttle. Carmen will remain in the shuttle and monitor the Void from the scanner."

"All right, good luck! Don't forget to televise your first steps on Titan, as requested by headquarters." This was an historic event for many folks back home. Garibaldi had linked up the video with a relay transmitter on *Deep Explorer* so that viewers on Earth could watch live the first steps of a human onto the surface of Titan.

"Sure, the onboard digital monitor is already recording. Carmen will videotape us when we step out and we'll take a portable recorder with us so you can see what we see."

After all their prepping they hit a snag when they had to open the sliding exterior door, which seemed to be jammed. After several unsuccessful tries to get it to open, Joe offered a suggestion. "I think the problem might be that some organic goo liquefied on the outer hall during our descent that glued the door shut or froze it shut."

"If we divert some heat to the outside we should be able to melt or vaporize it." Kitahari suggested.

"Excellent idea. Let's do it!" Zadak said.

Pablo rerouted the heating to the door. "That should take care of it in a few minutes," he said with confidence. Indeed, after a short while they were able to open the door. Kitahari extended the ramp and Carmen filmed while Zadak and Pablo were walking down the ramp.

When they opened the door a playa-like landscape extended in front of them. It was flat and smooth with the exception of some smaller rock and ice clumps interspersed within the fine matrix. A gentle wind was blowing over the surface. Zadak walked down the ramp first, followed by Pablo. It was unclear what the cause was. Perhaps, the ramp was at a too steep of an angle, or maybe Titan's organic goo underneath the ramp wobbled, or maybe the nervousness that Zadak felt at being in the spotlight, or some combination of these. Either way, when Zadak took his first steps down the ramp he slipped, his body weight shifted backwards and arms flailed in an attempt to catch himself. He would have smacked his helmet-covered head onto the edge of the ramp if Pablo would have not reacted instantaneously, catching him and reducing the impact.

Kitahari rushed toward them and both checked whether Zadak was okay. Zadak was swearing something in his native Serbo-Croatian that nobody understood as he tried to rise to his feet. Zadak held his back and waived to Pablo to move on. The only thing hurt was his self-image. Pablo and Kitahari walked a few steps further down the ramp, and stepped off into some slippery organic goo on top of ammonia-water ice bedrock. The goo was a few centimeters in thickness. It had a soft consistency and spread out as soon as they stepped on it. Although their suits were insulated very well, they felt the cold creeping up from the ground. They took each others' hands to get more stable, and Carmen finished her live-video with the words:

*"Together, hand in hand, the first woman and man stepped onto Titan, the first planetary body in the outer solar system on which humans will leave their footprints."*

Pablo and Kitahari went back on the ramp and tried to help Zadak back to the shuttle, but he insisted that he was fine and wanted to step on Titan's surface. They helped him and together they took a few more steps in front of the shuttle. Carmen stayed inside the shuttle monitoring their walk, and used the scanner to determine the position of the Void.

Jack and the rest of the crew on *Deep Explorer* watched every step the landing team took. Zadak's misstep reminded Jack of the story of King Harold that he once heard, who supposedly had slipped stepping onto the shores of England when he was landing with his army. His troops were very superstitious, and that surely contributed to his death and loss of the following battle over Britain. This misstep here should be much less dramatic, although Jack was sure that it would not sit well with the General who was now in charge of their mission. They weren't exactly putting on the best show for primetime coverage. In some way, though, he was pleased that Kitahari and Pablo were the first humans on Titan. He enjoyed their youth and enthusiasm. It was a stately symbol of humanity, companionship, exploration, and achievement.

After much finagling Carmen was finally able to pinpoint the position of the Void as being 186.7 m south, just underneath an ice ridge. Given the difficulties of walking in the goo, Zadak decided to fly to the site rather than walking or sliding to it. Kitahari, Pablo, and Zadak climbed back into the shuttle and Carmen flew them to the Void's location. Zadak didn't observe any obvious signs of the Void that would distinguish its position. It didn't look much different from their previous location in any special way. Just a few more and larger ice boulders were scattered on top of the goo-ice mixture. After a couple of missed approaches Carmen was able to land the spacecraft only a few meters from the Void's location.

"Anything special associated with the location of the Void?" Jack asked through the communications link.

"No," Carmen answered. "The same as with the Void on Mars and Venus. Nothing special would reveal its position in the topography—or even its existence at all—if we didn't have our scanners."

The landing crew exited the spacecraft and laid out a grid of millimeter accuracy over an area 5 m long by 5 m wide. Then they put a scanner on a tripod, above the position of the Void.

According to their measurements the Void was only a few centimeters below the surface, possibly directly below the 2 cm thick layer of organic goo. They unloaded a power generator and a communication relay to network their scanning equipment, intending to record and transmit every little energy burp from the Void to *Deep Explorer's* central computer. They also attached a video camera to the communication relay, which was moving one degree per minute over the immediate area of the Void. Afterwards they collected several ice and rock samples from the surface for analysis. They re-boarded the shuttle and flew above the immediate vicinity of the Void. The landscape was eerie with flow features carved into the organic goo. It reminded Joe of dirty ice on Earth.

"Do you see anything that could indicate life?" Pablo asked.

"Not that I can make out," Kitahari responded.

After a few circles over the landing site, they flew back to *Deep Explorer*.

Jack was happy to have his landing party back in one piece and met them when they got back onboard the ship.

"Sorry to have screwed up our live-broadcast," Zadak grumbled when he entered *Deep Explorer*.

"No worry!" Jack responded, smiling. "That can happen to anyone. The important thing is that the mission was safe and successful. We have our grid and should get very accurate readings."

## 4.3 Anxieties

An hour after the return of the shuttle, they had a meeting in the conference room. Zadak reported on the mission and concluded that with the exception of him falling on the ramp the mission was a complete success.

"What's next?" Kitahari asked.

"We will stay in orbit for the next few days monitoring the Void and analyzing the samples we brought in," Zadak responded. "We may have to return to the surface to modify some of scanning equipment, but most of our set up is complete."

After the end of the meeting Jack announced to everyone that Mike Hang had been replaced as NASA administrator by General Bicker. None of the crew seemed to be happy about the latest turn of events on Earth. For the most part everyone greeted the news with shock, misunderstanding and concern.

Zadak took the news in an unexpected manner, though. He just shrugged his shoulders and wordlessly flipped through the initial scans of the Void as if completely unsurprised by the turn of events. He seemed more annoyed at the distraction than anything as everyone else argued about how this change of events would affect them.

Garibaldi was the most upset. "If they think we are now their little toy soldiers…"

Jack interrupted him. "None of us is happy about this, but it probably won't affect us here at all."

"Let's hope so." Carmen said.

Garibaldi growled in disgust. "How could it not affect us? If there wasn't anything to worry about then there never would have been a need to replace Mike Hang."

Jack couldn't dismiss their fears but maybe he could alter their perception. "If the aliens are peaceful then we have little to worry about and I'm convinced things back home will return to 'normal' once we've assuaged most of their fears. But of course if the aliens are hostile…" he left the conclusion up to his listeners.

"Then we will probably all be killed." Zadak concluded for everyone in a nonchalant scientific matter-of-fact manner. "Either that or captured and tortured as they probe the strengths and weaknesses of our species."

Jack strengthened to his full height. "That may be the case, but either way the people back home have to know the answer and it does us no good to dwell on the possibilities and let our fears guide our actions. If we did that we would be no better than the bureaucrats back home."

Jack's final comment cast a calming net over the crew so Jack moved on. "Any new insights about our upcoming "Close Encounter of the Third Kind"?—referencing a famous old "first-encounter" sci-fi movie. He turned to Vladimir.

"No, not really. We get better and better resolutions, but nothing new. Our encounter will still be in about five weeks assuming constant velocity of the alien planetoid and assuming we stay here in the Saturnian system for the encounter."

"If we would leave here that would be totally stupid," Garibaldi said. "Saturn and its moons will give us at least some cover if something should go wrong."

Zadak and Carmen agreed, but Joe had his doubts. "They are probably already well aware of our presence and position and are probably monitoring us to see what we are doing. They certainly aren't entering our solar system blind. They probably already know a great deal about our society, culture, capabilities and technological advancement. To them it might appear as if we are hiding in the shadow of Titan and planning to ambush them. This ship does have weapons, after all."

"If we were to move toward them now, then we would buy Earth some more time," Chan proposed. "It may also have a psychologically stronger effect on the aliens; showing them that we are not afraid, and are willing to directly face them."

"That doesn't help us if we are blown to pieces!" Kitahari argued.

"If that is their intent, then it doesn't matter really where we're blown up," Chan countered coolly. "We are still destined to face them, after all."

"I think it would be more prudent to monitor them as they approach us," Vladimir suggested. "This way we gain as much information as we can and are prepared as well as possible for the encounter. There are no guarantees, I agree, but the more we know about them, their ships, and the Voids, both their Void and the one on Titan, the better for us. Besides, who knows whether they have the same type of psychology we do? Maybe for them we are only a stupid little fish who is curious about the big white shark.

It was silent for a few moments.

"I agree." Jack said. "Let's do it like Vladimir suggested. Vladimir, you continue to monitor the KBO, while the rest of us continue to work on Titan under Zadak's lead."

The crew dispersed, only Chan stayed a bit longer.

"You should be aware that there is an increasing amount of anxiety in the crew," Chan said to Jack.

"Of course, there is. I felt it. It's certainly understandable. We will have a historic encounter, a first encounter with an unknown alien intelligence. People tend to fear the unknown. And many people respond to fear with aggression."

"But fear will cause mistakes, mistakes that can be fatal."

"Yes, but we should not jump the gun either. We'll wait and let them come to us. We will meet them here in the Saturnian system. We won't let them pass by without contact. If they are hostile, the Saturnian system will be humankind's first stand, and if that costs us our lives, so be it." Jack spoke with conviction and without fear of death for himself. He was more concerned with his crew and beyond that, the people back on Earth. "I would like you to intensify your work on encounter strategies and scenarios and any options we have. Any theories you might have on protocols for initiating first contact would be quite useful."

Chan nodded and went to his room. Jack looked for Vladimir.

"Anything else?" Vladimir greeted him.

"Yes, I need you and Chan to develop a plan of encounter for the aliens and develop options for various contingencies."

"All right. You got it. When do you want to have it done?"

"Yesterday!" Jack grinned, but amended himself. "In about three days. I would like to see a sane plan generated by us before Earth comes up with their own plan."

Then Jack went to his room to be alone. Finally, he had some time for the in-depth meditation he was looking forward to since arriving in the Saturnian system. He shut the door of his room with a sign "Only Disturb in Emergency", opened a box with scent that he had received from Master Lao, and went into meditation posture. He let his mind clear and focused on his breathing. He felt the calming warmth moving through his abdomen and chest. Then he tried to open his mind and let his thoughts drift towards the Voids and the alien planetoid. Perhaps there was a chance to get into spiritual contact with the Void. His scientific mind perceived the idea as ridiculous, but perhaps there was a chance, as Lao had suggested. Lao had showed him many things he couldn't fathom with his scientific analytical mind. Jack's job was to find solutions that others couldn't find, and extraordinary circumstances might just call for extrasensory tactics, so he had to explore every possible avenue. And like it or not, he thought, the near future of humankind might depend on him and his approach. Jack imagined the Void, the empty space of nothing, and tried to connect to it, moving in his imagination towards it to unify with it. His respiration increased and his increased blood circulation made him feel warmer. His feet were falling asleep, but he ignored them, and continued his quest. At times he felt dizzy and trembled. He seemed to be able to access his deepest interior, but only his own. His joints and muscles were beginning to hurt, but he ignored them. Finally, he changed position and lay down on his back with arms and legs stretched apart in a shape of a pentagon, the old "Vitruvian Man", with equal distances between his arms, legs, and his head. His body, now at rest, was not hurting anymore. Perhaps, he thought, he could now better focus on the Void. His mind and spirit relaxed further. In his mind, the imagination of the Void as an empty space changed to a bright yellowish light, like the biggest star he ever saw, perhaps ten times the size of Venus at its closest encounter with Earth. It felt good and comforting, but there was no communication, at least nothing tangible. Then he let his mind wander to the Void on the planetoid and the intelligent life on it. It did not appear to him in the yellowish light but remained dark and empty. He tried it again. At some point he must have lost consciousness but when he woke up, he sensed a diffuse, roughly human shaped being standing in front of him exerting a dark, heavy force onto his chest, apparently trying to stop him from breathing. He gulped for air and struggled to bring himself fully awake and sit up. As he awoke the sensation and the force diffused.

Jack got up, breathing heavily. He was uneasy about the event and couldn't shake the thought of the strange experience. His mind was wondering whether it was real or only a bad dream. As far as he could recall he had never had a dream like that. And he hadn't intended to actually fall asleep. He must have been exhausted. The last days had been quite wearing. But it had appeared so real! His training at the Complex taught him not to dismiss feelings and experiences that had no obvious "scientific" explanation. This experience, whatever it was, would have no affect on how he dealt with the upcoming challenges, he vowed. If anything, he thought, it only confirmed his previous attitude that the Voids were no threat, but that he would have to deal with the alien planetoid cautiously.

## 4.4  Life on Titan

Jack opened his room door and strolled into the main laboratory, where there was a lot of excitement.

"What's going on?" he asked Carmen.

"Look. See the ice boulder on our millimeter grid?" she said pointing to a small rock on the imager. "We think it moved!"

"What?" Jack said incredulous.

"Yes, look, it was there yesterday, and today it's over here! It moved about five millimeters. That's not much, but it's significant. The movement is REAL!"

Carmen grew more and more excited. "We need to get down there and look at it. Or better, bring it aboard for study."

Jack turned to Zadak. "What do you think?"

"Carmen is right. It definitely moved. I have Joe and Pablo working on possible explanations such as ice-settling, solifluction, or even the wind.

"Oh, come on, Zadak, there is nothing physical that could move the boulder that way! It must be life-related!" Carmen argued.

"I'm not saying that you are wrong. I just want to exclude any other physical or chemical explanation," Zadak said looking at Jack.

"That seems prudent," Jack agreed.

Carmen went back to the computer screen and Jack continued his tour through the ship.

Everyone was busy with analyzing samples, monitoring the Void, or monitoring the alien planetoid. Jack made his scheduled telecon with HQ to report about the landing. Colonel Haggerty appeared and Jack was relieved that the unpleasant General was not available. However, the Colonel informed Jack that the General was very upset with the landing and felt that Zadak's performance made them all look like idiots. Jack wanted to say that from the fringe of space the military and all the leaders of Earth seemed like the bigger idiots, but he knew that kind of comment wouldn't help. Jack doubted whether the people on Earth really cared about him and his crew at all. Their only concern was the approaching planetoid.

The next morning Carmen and Zadak knocked on Jack's door and woke him up.

"What is going on?" Jack said. They entered his room.

"Carmen is right, that boulder is moving by itself!" Zadak said.

"It didn't seem to do much during the day when I continued monitoring it, but it moved a couple of millimeters during the night when we were sleeping. This time it moved in a different direction, basically eliminating all other non-biological explanations. It also changed its shape slightly. Look, that was yesterday morning," she said pointing to an image on a grid, "and this is today."

Carmen held the two images in front of Jack's face. He wasn't quite awake yet.

"Okay, okay, I'll go down with you," Jack said.

"Jack, you should stay on the ship!" Zadak said. "Someone else can go."

"I want to see some of the action, too, you know. Besides, there is nothing special going on on-board right now. You are in charge until I'm back."

"Come on, let's get suited up!" Jack said to Carmen as he yawned to clear out the last of the sleep haze.

They went to the changing chamber and got the shuttle ready. "You know, Zadak is right. You should stay on board," Carmen said.

"Perhaps, but I have to get out. Besides, this gives us the chance to spend some time with each other." Jack tensed after he said that last statement. He hadn't meant it to come out quite the way it did. Jack had respected their agreement to maintain a professional agreement and didn't want to jeopardize the close working relationship that had evolved.

Carmen smiled and if she had any problems with just the two of them going to the surface she didn't mention it. Jack was relieved, but silently cautioned himself to maintain his professional integrity. It was enough that they were alone together. He could revel in that at least.

Jack detached their shuttle from *Deep Explorer*, and initiated the thrusters as they descended into the atmosphere. Jack's stomach lurched as the g-forces rose and the whole shuttle shook and rattled like a runaway train. He had never liked atmospheric entries very much.

They set the shuttle down a short distance from the grid. They were mindful of the boulder-strewn plain they landed on and tried to avoid disturbing the boulders in the event that Carmen's observations proved correct. Then they climbed out of the shuttle and moved slowly to the ice boulder, being careful not to disturb the grid layout.

"Doesn't look very special to me," Jack said.

"Look here, the boulder has the shape of a slight arc in the direction of movement. It is not all the way connected to the ground. I would have noticed that, if it was that way before. It reminds me of how amoeba move. They extend part of their body forward, and then drag the hind part up towards the front. It is a very efficient way to move if you don't have any feet or arms. Do you see any drag marks in the sediment?"

"Yeah, there are some slight indentations, very gentle, that appear to vanish quickly in the viscous goo. Let's see whether we find some more of those arch-structured ice boulders in the area here."

They indeed found several others, and all had the characteristic indentations close-by. "Alive ice-rocks? None of this really proves anything. Maybe it's a physical or chemical phenomenon," Jack continued.

Carmen took a little heater and evaporated some of the ice crust on the arch-shaped ice boulder. Then she went over it with a scanner. "Organic, like I thought!" she said.

"Well, let's take a few of those with us. They won't fit in one of our sampling containers but we can just pack it in a lot of this ammonia-water ice to keep it cool. Zadak are you watching us?"

"Yes, Zadak here," he replied.

"Prepare a large cold compartment on board so we can study our specimen on the ship—say at about –180 C."

"All right. Are you sure you want to bring these things on board? Or should we get the analysis equipment down to you?"

"No, we can get them up to the ship. But we should keep them in a separate quarantined compartment."

"If they are truly alive we have no idea what sort of defenses they might have."

"Defenses?" Jack asked. "It's just an organic rock."

"Every living thing has some kind of defenses, passive or aggressive, to protect it from outside threats, both internal and external. I don't imagine this 'rock' is going to bite you, but we can't really imagine how it could respond to external stimuli. Perhaps it will ooze a toxic goo onto its surface. Either way, it is better to be prudent."

"You are right, Zadak," Jack responded and gave the rocks a more cautious scrutiny.

Carmen and Jack carried the boulders to the shuttle. One of the boulders alone would have weighed more than a hundred kilograms on Earth, but the lighter gravity on Titan made it easy to move them. Nevertheless, it was a challenge to carry the bulky objects on top of the slippery organic goo to the shuttle. Finally, despite being splattered and smeared with the organic mud, they were able to get them into the shuttle and packaged into a container previously used to house power cables for the generator at the Void. They packed them with a lot of ammonia-water ice and then quickly prepped the shuttle for launch.

When they arrived back at *Deep Explorer*, Zadak, Chan and Joe were already waiting and helped to carry the specimens to a special compartment in the lab room that Zadak and Garibaldi had hastily set up.

The science group lead by Zadak and Carmen went to work analyzing the specimens. They were able to cool the compartment down to –170°C, close to Titan's surface conditions.

After securing the sample in the science lab Jack went to Vladimir in an upbeat mood. Vladimir turned to him and smiled, "I hear you had a nice field trip."

"Yep. It was very enjoyable. It was really nice to finally get out of this can. Anything new while I was gone?"

"No, but Earth headquarters left a message that they would like to address the whole crew at 2200."

"Uh, oh! New trouble?"

"Probably, but tell me about your Titan adventure. Did you find alien life?"

"We think so!" Jack sat down and told him about the details of their trip to the surface. Vladimir was as amused as usual and commented that perhaps Jack could name the living rock after himself; a pair of similar hard-headed and slow creatures.

Jack called for a meeting at 2000 and everyone assembled in the conference room on time except Carmen, who was ten minutes late. Jack began as she walked into the room. "I called you together because we have a major scientific breakthrough. Carmen, would you please enlighten us on the latest insights about your ice creature?"

"Sure," she said. "I'm delighted. It is actually not an ice creature. It uses ammonia-water ice only for insulation. Beneath its crust or exoskeleton, it is all organic. It's more like a slug if anything. It moves along the surface and devours anything organic it can find. The major metabolic pathway seems to be the oxygenation of acetylene that is created in Titan's atmosphere from photochemistry. The acetylene falls down to the surface in solid form due to its greater specific gravity."

"Isn't acetylene an explosive?" Kitahari inquired.

"Yes." Carmen continued. "Under Earth conditions it is. But here on Titan, the environment is so cold that it is very stable. You need something with lots of internal energy to drive metabolic reactions here. The slugs may also be able to use a lot of other organic compounds, anything from which they can extract energy. Even radical chemical compounds that are created in the atmosphere by the interaction of UV radiation and the organic haze are used. It looks like they combine some of those radical compounds to build hydrocyanic acid and cyanamide. Cyanamide, in turn is then used to build organic macromolecules. It glues amino acids together to build proteins. Or so I think."

"Wow, so they use the same biochemistry as Earth life?" Lana inquired.

"No, not really. We haven't found any DNA or RNA. They must use some kind of other chemical code. We suspect their DNA equivalent has a polycationic backbone rather than a polyanionic backbone, which is used by Earth life. We've found some matching chemicals in the creature, but we have to further investigate whether that chemical molecule is really their information code."

"Also, they use silicon-rich compounds, silanes, as Earth life does lipid molecules," Zadak added. "An ideal adaptation to the Titan environment and not seen in any known Earth-type organism."

"So, we truly have life of a separate genesis here on Titan?" Jack asked.

"It appears so." Zadak continued. "But we can't be a hundred percent certain yet."

"The organism can move, right?" Joe asked.

"Yes, very similar to amoebae on Earth," Carmen took back the initiative. "It can take practically any shape. One part or appendix of these amoeba-type creatures extends to where they want to go. Then it drags the rest of the body after it, extends again, and so on.

"How do they eat, then?" Joe asked again.

"It appears that they just engulf their food —again, like Earth's amoebae. We can't find any specific tissues or anything that would resemble an organ. Their metabolism is really more like that of an oversized microbe than that of a slug, I guess."

"For their interior workings that is probably true," Zadak added. "We posted a note and a science breaking news release, where we referred to them as "Titan slugs.""

"That should have been cleared with all of us," Chan noted. "What if it turns out that you are wrong?"

"Chan is right." Jack noted and turned to Zadak and Carmen.

"It is not a publication or anything!" Zadak defended. "Just a transmitted news note to the press. We figure that anything that can show our mission in a positive light and shifts the focus back to scientific discovery and away from that alien planetoid is worthwhile. We wanted to make sure that the news is out before our 2200 conference call with Earth."

"What is the call about?" Chan asked.

"I really don't know." Jack answered. "I was not informed, but I don't have a good feeling." Jack wanted to say more. He was not very fond of that General, who seemed to have his own agenda and didn't give a damn about Jack, his crew, or their science objectives. But as the commander of the mission it would be irresponsible for him to portray a lack of confidence in the man in charge. That would be asking for trouble even if Jack's misgivings were completely accurate. "We will route the transmission to this conference room so that everyone can join," Jack continued. "Pablo, can you take care of that?"

"Sure thing!" Pablo responded as he yawned and stretched languidly.

"All right, see you in one and a half hours."

## 4.5  Rebellion

Everyone was in the conference room when the transmission came through, with General Bicker on the main screen and Colonel Haggerty next to him. The expression on the General's face gave Jack's guts a twist. Nevertheless, Jack kept his calm and started the conversation. "Good evening, General. I hope you had time to read our transmission about the discovery of life on Titan."

"Ah, yes," the General responded. "We will come to that issue later. People here wouldn't give a damn about your discovery even if you found Titan crawling with amphibians. We are much more concerned about national and world security right now and not about an academic wild-goose chase. Our remote sensing results indicate that the alien planetoid has extensive constructions on the surface, and at least twenty-five orbiting ships. Dr. Kulik, does that concur with your findings?"

"That is correct, Sir. Most of the constructions appear to be tunnels to the subsurface," Vladimir responded.

"That alien planetoid is our prime concern," General Bicker said again as he continued. "This thing is on a direct course toward Earth and the President's advisors and I agree—we all believe our best shot to defend our planet is to destroy their Void."

"We don't even know if they are hostile!" Carmen interjected. "And this is an international mission. What does the world council say?"

The General simply ignored Carmen's comment and continued. "Our scientists have figured out a way to destroy a Void. They believe it can be done with the same kind of laser you have on board your ship. You just have to rotate the frequencies and fire the laser in specifically tuned pulses."

"What do you mean 'they believe'?" Jack asked in a passive voice, though his anger was threatening to boil over. "It didn't work well on Venus… or whatever Dr. Ludmilla tried to do there!"

"Of course, it has to be tested first. You are going to test it on the Titan Void first to see if it works."

The conference room erupted into a cacophony of shouts, rejections and disbelief. Jack turned toward the crew and tried to calm them down. They fell silent but the shock was still plain on their faces. Zadak gritted his teeth and clenched his fists as he fought to contain his anger. Jack watched as his face went from white to red to red-purple in quick succession and he wasn't sure if the military-phobic scientist would be able to contain himself now that his worst fears seemed to be coming real. Jack gave him a stern look to help calm him down before he turned back to the General.

"So you want us to destroy the Void on Titan and possibly with it all life that we just have discovered? And to do this based on a possibility of hostile intent and threat from the alien asteroid?" Jack inquired.

"Exactly!" the General said coldly. "But let me explain the position to all of you in plain terms, since you don't seem to have a grasp of the gravity of the situation. First, we don't know that the Voids are really connected to life. Second, and more importantly, the alien planetoid is a grave threat to humanity and this approach is our best chance to defuse it without major damage to Earth."

The conference room erupted again into a chaos of raised voices, harsh words and calls for rash actions. Jack sat back for a second to let the rush of thoughts and emotions settle. He only dimly recognized the voices of the others and their pleas for reason and leadership. Only moments passed but it seemed like an eternity to him. The image of the yellowish light of the Void resurfaced in his mind.

Finally, Jack said, "I can't do that, General Bicker. It is against all that I believe in. We are here to discover and preserve life, not to destroy it."

The General's face remained a cold emotionless mask, but it flushed a few shades redder. "Is that your final response, Dr. Kenton?" The menace in his voice was so intense Jack wanted to turn away from those cold eyes. The conference room fell so quiet that Jack suddenly became painfully aware that all eyes were on him. The 'gravity of the situation' as the General had described it, couldn't get any more obvious to Jack. This wasn't a ball game or an exercise where he could break the rules without any consequences. He had been given a direct order, essentially from the President of the United States, which if followed could possibly destroy all life on Titan. Did anyone have the right to drive a species to extinction on a whim with so little substance as this merely presumed threat? And beyond that, he was being ordered to pursue a course of action that would inevitably result in either the annihilation of the alien life forms or the possible destruction of every human on Earth. These other aliens could be millions of years beyond the human race technologically, and whether the aliens were hostile or not, attacking them could be tantamount to planetary suicide. Furthermore, if there was going to be a war with the aliens, he sure as hell wasn't going to be the one to start it and doom his entire species to extinction.

Jack met the General's gaze with all the strength, courage and conviction he could muster. "Yes, General, that is my final response."

"Then, I'm forced to relieve you of command of this mission, and hand over command to the second in charge, Dr. Chan."

"Dr. Chan," the General continued. "You will adjust the laser frequency, destroy the Titan Void, and break orbit to intercept the alien planetoid. Is that understood?"

Chan felt all eyes upon him and he turned a shade paler than usual. There was a surprised and hushed pall in the room. Jack was stunned by the revelation and turn of events. Panic shot through him as the possibilities popped into his imagination. Why hadn't he put Garibaldi or Zadak officially as second in command and informed headquarters? He had wanted to do so, now it was too late. He'd simply forgotten, or thought there was never going to be an urgent reason. It was sloppy, an inexcusable mistake for which he surely had to pay now.

Chan swallowed hard and looked furtively from Jack to the crew and then back to the general. Everyone held their breath to see what Chan would say, and the silence threatened to overwhelm Jack's fragile grasp of self control.

"No, Sir. I will not!" Chan finally answered firmly. "You do not have the authority to replace Dr. Kenton. He remains the leader of the mission. Over and out," and Chan turned off the transmission.

Jack nodded his thanks to Chan. Zadak clapped Chan on the shoulder. The others let out a huge sigh of collective relief and then applauded briefly—the roomful of tension had completely evaporated.

"We have to figure out what to do now," Zadak said, interrupting the rejoicing.

"They need us more than we need them." Garibaldi added. "They have no real power over us. They will come around and be more reasonable, I believe. And if they don't, SO WHAT?"

The crew continued to discuss the implications and issues involved in their increasingly perilous situation. Thoughts flew back and forth and emotions rose and ebbed, but after an hour of deliberation Jack put a stop to the discussion.

"We proceed as planned!" he said resolutely. "We obtain as much information as possible before taking any actions. Zadak, you will lead the science team studying the Void, as before. Chan and Vladimir, you will continue to work on encounter strategies. Carmen, you will head the investigation of the slugs."

The crew went to perform their assigned tasks, but Jack grabbed Chan by his sleeve as he was leaving. "I want to personally thank you."

"It was the best decision for the mission," Chan said coolly. "Everyone views you as the mission commander; not me. If I had accepted command we would be left with infighting and would not be able to focus on the upcoming challenges. I hope you are right, Jack, and that you chose your approach not because you are unable to make the tough decisions. We both know, or feel deep inside, that those coming aliens are hostile."

"But if we destroy all life on Titan, on a whole planet, we will be worse than them and whatever we stand for will be a farce." Jack countered.

"Perhaps, but then that life seems to be only consisting of big microbes on a moon far away and we may have failed to take the only measure that could have saved Earth. Your action may turn out to be too idealistic."

Chan was about to leave but then he added with a smile. "Isn't it interesting how quickly we are able to reject authority and embrace independent decision-making the further we are from Earth? I don't think either one of us would have been so bold if we would have stood right next to the General."

Jack nodded his head in agreement. "You're right, Chan. Fear of reprisal or consequences becomes a lesser issue."

Then Chan left to find Vladimir, leaving Jack alone in the conference room.

Jack reflected on Chan's words. Perhaps he was wrong. But it just didn't feel right to destroy the Void on Titan, and possibly all the life on Titan with it. He would have felt like a mass murderer, even if he would have killed only big microbes. Everything felt wrong about it. But what if that really were the lesser of two evils? Then, he guessed, the 'logical' choice would be to make this smaller sacrifice to save Earth. But still it didn't seem right. Chan was correct about one thing though. He would have to live—or perhaps die- with the consequences of his decision if he was wrong.

## 4.6  Alone in Space

Zadak decided to collect more samples from the surface and established a small Titan Biosphere II within the main laboratory on *Deep Explorer*. Zadak and Pablo worked on analyzing the pulsating pattern of Titan's Void and compared it to the other Voids' patterns. At one time they needed an updated pattern of Earth's Void, and requested it from HQ. After a ten hour delay they received the data. Vladimir worked on the pulsating pattern and energy signals from the planetoid's Void and befriended Zadak over a bottle of home-brew he'd been able to secretly brew up. This way, Vladimir obtained easily Zadak's assistance to analyze the mountain of data he had collected. When comparing the signals from the various Voids, it appeared that the timing of the random patterns within the pulses was a function of the relatedness of the Voids. While the time intervals were nearly identical for Venus, Earth, and Mars, they were offset for Titan and even much more offset for the KBO. While Vladimir and Zadak were absorbed with that project, Jack worked with Chan to come up with a plan for first contact.

Three days had passed without any direct contact with Earth and Jack was getting concerned about the crew's reaction to their defiance. He still transmitted his daily reports, but received back only an acknowledgement of receipt. He tried to get in contact with the Complex, but this proved to be unsuccessful. Jack had no idea why but figured that either the Complex's frequencies had been changed or his transmissions were being blocked. Everybody was busy doing their work and nobody talked much about their lost connection with Earth or the upcoming alien encounter. It was a strange atmosphere on the ship. They felt like a group of renegades committed to go forward in what they believed in—no matter what the consequences.

"I have never felt so disconnected in space before," Carmen confided to Jack once when he was sending his daily report back to Earth. He nodded agreement to reassure her. He felt lonely too, and the burden of responsibility fell heaviest on him. Everyone was looking up to him to decide what to do to save them and Earth. If Earth turned out to need saving, that is.

A week passed by. Then, finally, Earth requested a telecon.

The entire crew was present again at the prearranged meeting time. When the call finally came through and Colonel Haggerty's image blipped onto the screen, the Colonel took the initiative right away.

"Hello, Dr. Kenton. We apologize for leaving you in limbo after our last formal telecon. There have been some changes in command here. General Bicker has been re-assigned to oversee construction of the space ship fleet. I'm in charge of this project now and I was able to convince the President to re-establish contact with you. I am not a diplomat, Dr. Kenton, so I'll be frank. There has been much discussion here concerning your loyalty and intentions. I would like to discern what those are and listen to what you have to say. Since I know the majority of you are not insane revolutionaries, I'll assume a greater part of you want to be team players. I also brought some friends of yours to demonstrate my sincerity."

The view widened and Mike Hang and Dahai appeared next to the Colonel.

"Hi, Jack!" they greeted him.

Jack's face relaxed and opened up. He wasn't sure about Haggerty's intentions or sincerity, but he was glad to see a couple of people he trusted much more than he did the Colonel. But that might be exactly Haggerty's ploy. Geez, he really was beginning to think like Zadak. "Great to see you. It can get pretty lonely out here."

"I bet!" Mike responded. "Finally, some of the hardliners around here saw some reason and they would like to assist you in your upcoming alien encounter as much as they can."

"Are you back in charge?" Jack asked dubiously. He agreed with what Garibaldi had said. It seemed headquarters was more willing to compromise now that they had realized the crew wouldn't blindly follow orders. They were scientists after all, not Marines.

"No, my status hasn't changed. I still serve as advisor to Colonel Haggerty. But he and we all are committed of helping you in your mission."

"Has he been taking your advice, Mike?" Jack boldly asked in a seemingly friendly tone.

"He listens, Jack. The fact that you and I are even having a conversation is proof of that."

Jack was glad to hear that tidbit and so were most of the crew. Colonel Haggerty took the break in conversation to get back on topic. "Do you have any plans for the alien encounter, Dr. Kenton?"

"Yes. I'll send them to you shortly, and you can give us your input. What we really need is any update or any insight that you might have gained on what caused the void on Venus to explode. If the aliens are hostile, we probably don't stand a chance in combat. We need to understand the Voids. We believe they have some greater meaning and are the key to opening up more possibilities. That sort of knowledge seems to be the only thing that might give us a chance."

"We came to the same conclusion," the Colonel said.

"We also made quite a bit of progress on the biochemistry of the Titan slugs and the emission patterns of the Void on Titan. My crew is working around the clock. We would like to send you our results and receive your input. We feel the better our understanding of these Voids and their relationship to life, the better chances we will have with the aliens. Perhaps, we can even use them for communication with the aliens, and avoid any potential confrontation."

"Absolutely right!" Dahai added. "We have to be non-aggressive."

"But assertive!" the Colonel added. "We can't appear weak or defenseless. You cannot let them pass by without making contact, Dr. Kenton. We have to know down here on Earth what to expect. I won't lie to you or your crew. I may very well be sending you right to your deaths, but that may be your task and also your duty."

"We understand." Jack responded.

Jack switched off the telecon and the crew rejoiced in relief. The human race wasn't going to hell in a hand basket after all and they still had some support and friends left on Earth. The stress, anxiety and feelings of isolation slipped away and were replaced by a renewed conviction about their mission.

It was still perfectly obvious that Earth wouldn't be able to help them a lot in their upcoming encounter. They were still on their own hundreds of millions of miles away from Earth, but just the feeling of being reconnected and supported by Earth made a big difference.

## 4.7 Preparations

The crew worked feverishly in the weeks that followed. Carmen lived in her own world with her Titan slugs, probing their biochemistry from all conceivable angles. She was supported in the lab by Joe and Kitahari, and Zadak, who made several trips to the surface to collect more Titan slugs for their Titan Biosphere II. All in all, things progressed well.

Zadak and Vladimir tried to understand the pulsating emissions. Amazingly, language algorithms gave the best matches to the emission patterns. It was not a 100 % match, of course, but it was better than the "ant reproduction cycle" match that they'd come up with on Mars. They reached a match of 87.3 % here. Both scientists, however, were doubtful whether the patterns actually had anything to do with language or communication.

Matter was vanishing and being replaced in perfect order and symmetry within the Void on Titan, just as on Mars and Venus. They confirmed this with various particle matter experiments. They had endless discussions about possible functions of the Voids and why they were there in the first place. The Voids seemed to be something from another world—another universe, really—and defied the conventional physics of our own universe. There were heated discussions. Some crew members like Chan and Zadak felt that they must be connected with the evolution of the Universe and its chemical and physical forces, while Kitahari, Vladimir, and Garibaldi argued they were too sophisticated and that they could have only been placed here by an intelligent life form. Carmen and Pablo felt uncomfortable with an intelligent design notion, especially if it was related to life, but didn't feel convinced either way. Lana and Joe felt that there were more important issues to discuss than the philosophical implications of the Voids. Jack offered no opinion but weighed the arguments as best he could and tried to keep considering all ideas as viable options. Vladimir liked the comparison of the Voids to mini-black holes, but he wasn't able to explain why the Voids did not impose any gravitational force at all. Zadak liked to compare them to mini-wormholes, as pathways to possibly other dimensions, but that still failed to explain how the Voids reorganized matter.

Pablo, Garibaldi, and Lana were working on upgrading the ship and the shuttles as much as possible. They improved their maneuvering abilities by streamlining thrusters and primary propulsion. In addition, they worked on the so-called Improvisation Plans designed by Jack and Chan in case certain parts of the ship became inoperative, either by hostile alien action or otherwise. They also considered improvised new weapons, none of which, though, sparked any excitement. If the aliens were hostile then a head to head conflict would surely be futile regardless of the firepower the Earthlings possessed.

Jack and Chan continued to work on encounter and first contact strategies. They also kept a close eye on morale, which was pretty good. Communications with Earth went smoothly and freely, now. HQ provided them with all the resources they wished for. In a strange way, they felt much closer to Earth out here than they had on their mission on Mars. Joe joked that was the case because they could watch their favorite TV series from here. But none of them really had time to watch TV anyway. Zadak cynically commented that they were being treated like condemned prisoners before the execution. The situation felt like the lull before the big storm.

*****

On Earth, meanwhile, large spaceship assembly facilities were hastily constructed. Nobody knew whether the aliens were indeed hostile, or if they would continue on their path toward Earth, or whether the laser cannon on board of *Deep Explorer*, or any other known weapons would have any effect at all. Nevertheless, it was felt the best strategy was to encounter any aliens from a position of strength and that they should be prepared in either case. Thus, resources were massed to build a fleet of *Deep Explorer*-like vessels to defend Earth, if necessary. General Bicker led the American construction effort. Nervousness amongst governmental leaders was at a high pitch as they realized that they had less than seven months left for preparation after the alien planetoid entered the Saturnian system.

Many voices criticized the lethargy in humanity's space exploration in the last nearly hundred years since the Apollo landings. They argued that a more proactive approach would have led humanity to possess better space technology and thus would have made mankind much better prepared for a possible confrontation. But, of course, that insight could not change the past.

*****

Three days until encounter: the crew on board *Deep Explorer* became increasingly nervous. Chan and Jack countered that nervousness by intentionally demonstrating relaxed confidence. Jack meditated more now than anytime since leaving Earth. He tried several times to connect with the Titan Void and he saw it again and again in a yellowish color, usually the size of an irregularly formed tennis ball, but he couldn't ever get any further than that. Once he tried to meditate and connect again with the alien planetoid, but he could not focus and let go. Apparently he was still too disturbed from his previous experience.

Chan proposed that they maneuver the ship in advance to an encounter position, but Jack rejected his suggestion and decided to stick to the plan. Vladimir provided detailed maps of the surface of the planetoid on which they could clearly distinguish details. Most structures were apparently underground, but details of the alien ships emerged. The larger ones were cigar-shaped, and some of the smaller ones were spherical. Vladimir could even pinpoint the Void on the alien rock within an uncertainty of a few kilometers. Contrary to the Void on Titan or any of the other Voids they had studied, the KBO's void was located deep inside the planetoid. Carmen finished the automatic setup of the Titan environmental conditions in the Titan Biosphere II lab, and joined as the last senior science member the AEG; the Alien Encounter Group, which included Jack as leader plus Chan, Zadak, Garibaldi, and Vladimir as advisors. Though they had thought up and reviewed dozens of hypothetical encounter scenarios, Jack wasn't excited about any of them. Whatever the final choice, and they had narrowed it down to three variations of one approach, their survival chances were slim if the aliens were hostile. The absolutely best thing they could do was to ensure that their first move was perfect. The main problem, still under discussion was how not to appear overtly hostile, whilst still being prepared for any eventuality.

Twelve hours before encounter: Jack was lying on his bed, relaxing and letting his mind filter through the possibilities, when he heard a hesitant knock on his door. Jack rose, opened the door, and saw Carmen. The look on her face was one of distress and perhaps something else Jack couldn't identify. "Is everything alright, Carmen?" he asked concerned.

"May I come in?" she asked softly.

Jack let her in and closed the door behind her. He wasn't sure what was wrong or why she was here. When she moved back towards him she was more confident. "I don't want to be alone, Jack. Not now." Jack felt a nervous tightening in his stomach, just like when he was a teenager and about to ask a girl for a dance. He didn't know how to respond or act, or even exactly what she meant by not wanting to be alone.

Carmen took him by the hand and guiding him to his bed. "Just hold me, Jack."

They lay down on the bed together, side by side, and Carmen wiggled under his arm and into a comfortable cuddle.

Jack drank in the feminine scent of her hair and the softness of her body. His imagination whirled with the possibilities and the wonderful sensations and emotions her touch and company gave him. It had been a very long time since he had held a woman in his arms and the excitement gave him a rush of pleasure, confidence and reassurance. He could die tomorrow and be a happy man—that was how she made him feel. He was relieved that she only wanted to be held. Anything more would seem cheap. No, all he needed was her touch and affection.

He laid there wordlessly, with Carmen in his arms, for some time, just breathing and silently reveling in the wonderful joy of intimate contact.

"Do you think we will still be alive tomorrow evening?" Carmen finally asked.

Jack took a deep breath and let it out slowly before he responded. "I don't know, but we won't go out without a bang. I don't want to be too negative about it. One way or another, it could also be the greatest moment in human history."

"Or the worst," she countered.

"True."

They lay silently next to each other again for a time. Finally, Jack said with a smile, "I think we should get some sleep before the encounter tomorrow. Remember I gave that as an order to the whole crew a few hours ago and you are one of them."

"I don't like to be alone, Jack! That's why I came here. But if you order me to leave then I will."

Jack suddenly felt like a jerk for even asking her to leave. He had no right to lord his authority over her at this time and in that manner. His responsibility was to the crew and right now she needed him. It would be wrong to turn his back on her in her moment of need. At least that was how he justified it in his own mind. He caressed her face and stared into her eyes to show his sincerity. "Then, stay here!" Jack held her very close and she fell asleep in his arms a little while later.

On the eve of which might be their last night, they were not the only ones that stayed close to each other. While Kitahari was searching for a candle, she saw Garibaldi and Lana vanishing together into Lana's compartment and caught a glimpse of Lana half undressed. Finally, after her search was successful, she went to the conference room, where Pablo and Joe were already waiting for her. They initiated a meditation ritual, where they held one-another's hands while sitting around a lit candle. Vladimir and Zadak were still working together analyzing the latest remote sensing results from the KBO, while Chan restlessly walked through the ship like an officer on his last inspection tour before battle. When he passed the conference room, he saw the meditation, or prayer circle—it wasn't clear what it exactly was. Kitahari invited him to join. Chan was hesitant and reluctant, but then finally accepted and joined in. It seemed that no one had followed Jack's orders and gone to sleep. It wasn't until very late in their usual sleeping cycle that everyone finally found some rest.

Four hours to the encounter. Jack had a final telecon with Earth HQ. It was really an exchange of best wishes and goodbyes. Jack felt that his counterparts were treating him like it was a last farewell. Not unreasonable, he thought: after all, what were the chances of them surviving the day? Fifty-fifty at best? They had all morbidly sent their wills home should the worst happen. After the telecon he tried to get in touch with the Complex and to his astonishment he got Lao on the other end.

"What a nice surprise!" Jack said.

"Well, I guess all other senior personnel are somewhere else, so at the moment I'm also the communication officer at the Complex."

"Are you posted on our strategy?"

"Yes, I am."

"What do you think about it?"

"It sounds reasonable."

"You know there is a good chance we won't be alive tomorrow. Headquarters is certainly treating us as if we're already dead."

"That does not matter. All that matters is what actually happens. Even if you do die, you will only unite with the spirit of the Universe. Is that not something to rejoice over? There is nothing to fear."

"Well, I know my crew is fearful. They are hiding it, but …

"Don't. There is nothing worse than fear. Be fearless, only without fear will you be able to master the encounter. Detach from the situation and connect with the Universe."

"I'm not sure I understand," Jack replied humbly.

"Don't feel trapped! You need to think bigger. Free yourself, and embrace the unknown; embrace the danger. Jack, you have it in you! You just have to listen to your inner self, the part that is connected to everything around you."

"Thank you, Lao," Jack said simply. "I have to go now."

"I know." Master Lao responded. "I'll be with you, always."

"Thank you." Jack said again and turned off the transmission. Then he went to the briefing room.

# 5 Deep Encounter

## 5.1 Countdown

They all met in the briefing room.

"You know what to do," Jack said solemnly. "We talked it over many times. We will proceed as planned."

All members of his crew nodded silently. Jack studied each of them carefully. They all looked tense, which was understandable under the circumstances, but they didn't look scared. They were resolute and strong. "If no one has anything to say then get to your posts and may God bless us all."

They all left of the briefing room scattering positive comments about success, courage and unity. A few moments later Garibaldi initiated the engine warm up sequence and Lana laid on an intercept course. Chan and Joe went to the shuttle and prepared for launch, while Zadak and Kitahari prepared the ice clipper shuttle for launch. Vladimir and Carmen monitored their approach to the planetoid on the high-resolution imager, and initiated the radio transmitter to broadcast their transmissions in multiple frequencies to try to communicate with the aliens. Pablo assisted Garibaldi with the propulsion system and ran continuous diagnostics to make sure everything was running smoothly. Jack was sitting as co-pilot beside Lana and was in close communication with each subgroup.

The next hour was a test of patience and self control. Each minute that passed only upped the tension on the ship another notch. Jack remembered his training at the Complex and worked to mentally purge himself of fear and doubt. That training would help them all and nothing could have prepared them better for this agonizing wait as they blasted towards total uncertainty.

Thirty minutes until intercept: they had not received any answer to their overt attempts at communication. They weren't that surprised, really—after all, Earth had been trying to contact the alien vessel for months to no avail.

"We have reached encounter velocity," Garibaldi announced from engineering.

"Are there any reactions from the aliens?" Jack asked the imaging group.

"Its not quite clear yet, but it seems that they moved about twenty ships well ahead of the planetoid with several others still orbiting it and shuttling to and fro," Vladimir responded.

"Okay, Chan and Zadak, launch your shuttles."

Chan and Zadak initiated the shuttles' thrusters and launched. Chan and Joe moved their shuttle to the right flank of *Deep Explorer*, slightly ahead of the mothership, while Zadak and Kitahari moved their ice clipper shuttle to the trailing left flank.

"I still think it is crazy to have an unarmed shuttle as the leading spacecraft of our ensemble," Lana commented.

"We had long discussions about it and Chan is aware of the dangers and agreed that this is the best course of action. A shuttle by itself is less threatening and if the worst should happen then we still have a chance to retreat. Besides, we don't know if our weapons would make a dent into those ships, if it comes to that. We stay the course as we agreed."

All systems on board the spacecraft were working at optimum. Their formation moved inexorably closer to the alien vessels. Jack ordered the shuttles to turn on all their video recorders and link them via a live-transmission to both *Deep Explorer* and Earth. Live-transmission was also initiated from *Deep Explorer* to Earth. Whatever happened within the next hour, at least Earth would have detailed records of it and might be able to respond. Even their possible destruction would give Earth knowledge of the alien's technology, tactics and defensive capability.

It's amazing what thoughts flirt through one's mind under such circumstances: Jack remembered when he had first faced death. He was fifteen years old at the time, but he remembered the experience well. It was a little lizard that had been driven over by a car speeding on an unpaved road. When he reached the street, the lizard lifted up his head and for a moment he felt this deep connection with the lizard. That was just before blood shot out of the lizard's mouth, and it collapsed and died. He asked Lao about that experience some months ago. Lao wasn't really sure about its meaning either but suggested that the lizard might have transferred his spirit to him during that encounter. Jack wasn't sure if he believed that explanation, but he couldn't otherwise explain the lifelong impact the memory had on him.

Now it was time to face death again; possibly his own.

Chan would be the first to intercept, only a few seconds before *Deep Explorer*. They were flying at the preplanned encounter velocity—slightly faster than the aliens' planetoid. They should be traveling fast enough to change direction, and escape if necessary. But they would only run if they were attacked. Otherwise, Jack wanted to hold the course, no matter what. At the very least he hoped to slow down the alien planetoid, if that was even possible. All his senses told him that this was a dangerous, hostile situation. But then he didn't want to be presumptuous. He hoped he was wrong. The high-power laser on board was ready for use, but only as a last resort. Even if he decided to use it, it would take at least thirty seconds to charge the laser's power cells for a discharge. Thirty seconds in a combat situation was an eternity. The laser was simply not designed for combat, but as a research tool, and its usefulness in a combat situation was doubtful. They had agreed not to charge up the laser before the encounter, because of the possibility that the aliens might be able to scan them and misinterpret it as showing hostile intent.

Zadak and Kitahari had their shuttle ready to launch a projectile if necessary. In order to do this they would have to move the shuttle in an awkward position, since the projectile was designed to be launched vertically, from the bottom of the shuttle, to impact on Titan's surface and then to collect ice for analysis. What an archaic weapon! Jack thought. It had all the sophistication and accuracy of a medieval catapult flinging a rock.

The alien ships were now well ahead of the planetoid with one ship at the very front, followed by four more ships about twenty seconds behind, and the third group of ships, about fifty of them, remaining close to the planetoid. It appeared that the aliens had moved most or all of their ships from their subsurface bases into space.

Apparently they don't know what to expect either. At least in that respect we are evenly matched, Jack thought. But why have they not attempted to contact us? Surely they must be capable. Or maybe they have tried, and we're so unsophisticated that we missed it? Surely they have been monitoring our television, radio and other broadcast signals since they became aware of our existence.

Vladimir reported that the sensors indicated the planetoid had slowed down a bit, while the lead ship was at the exact velocity of the planetoid a few minutes ago. This could confirm Zadak's hypothesis that this speed was the top velocity they could reach, Jack thought, which meant that *Deep Explorer* could move faster than they. Perhaps a glimmer of hope, at least they might have a chance to escape if necessary.

Vladimir and Carmen continued trying to initiate radio-communication targeting both the lead ship and the planetoid. Jack could now see the lead ship through the window as a bright star-like object, and he used the magnifier on the cockpit screen to make out its shape and some details. It had an egg-like shape and was about eight times the size of *Deep Explorer*. He couldn't make out any armaments, but then, he didn't know what they might look like, either.

Five minutes to intercept. There were still no answers to their transmissions.

"No changes in the direction or speed of their lead ship," Vladimir announced through the intercom. Jack began to get nervous. Somewhere deep inside he had still hoped for contact with the aliens before this encounter. But now he was moving closer and closer to what could be perceived by the aliens as a confrontation. And in a way it was, since they entered Earth's solar system and his duty was to stop them from advancing toward Earth. 'Their ships are much bigger and they have many more. But we shouldn't be perceived as a threat!' Jack thought. Either way, he had to continue with his course of action.

They could not slow down or stop. If they did, they would be sitting ducks. It all went through Jack's mind again and again, but fear of the possible outcomes would not be productive or helpful. That was the risk they had to take.

Two minutes to go. Chan's shuttle continued to advance further to the front. There were now a few seconds separation between Chan's shuttle, *Deep Explorer*, and Zadak's ice clipper. The aliens could not seriously consider the shuttle a major risk, Jack thought. It is like a mosquito approaching an elephant.

Lana watched and waited in the pilot's seat, seemingly cold and calm. She was ready to take evasive maneuvers at the first sign of trouble.

Vladimir informed Jack that the alien lead ship was slowing down, which he perceived as good news. Perhaps the aliens were ready for first contact. On the other hand, the following four alien ships were catching up to the leader quickly, and they would have to face five ships instead of one in short order. And there was still no answer to their radio transmissions. Jack didn't like the look of things.

## 5.2   First Contact

Chan's shuttle was less than a minute from the alien lead ship, when Jack saw a blue streak emerge from the alien ship, and speed toward the shuttle. It took only a fraction of a second to reach Chan's shuttle. When it did, it sliced right through the ship like a golf ball through a window. The air blasted out, scattering pieces of debris into space. The shuttle itself eerily continued on as if nothing had happened.

Jack sat motionless, stunned for a split second by the suddenness of the attack. "Damn it!" Lana shouted as she initiated maximum thrusters. *Deep Explorer* lurched forward as it accelerated.

"Chan, come in!" Jack screamed through the comlink, but there was no response. The video recordings inside the shuttle indicated complete chaos, hull breach, and vacuum inside. "Get us the hell out of here," he growled at Lana who was already vectoring in a course to evade the second wave of oncoming vessels. Jack felt nauseated. His worst fears had come true. "Shit!" he yelled as fear and rage rushed through him. But there was nothing he could do for Joe and Chan and the feeling of helplessness made the situation all the worse.

It was time to focus on the rest of the crew now. Lamenting his losses would not help the situation. "We'll pass the lead ship on the left side at full speed," he instructed Lana.

"I know what I'm doing," she shot back just as another blue streak emerged from the lead ship and hurtled toward them. Lana reacted almost instantly to evade the attack and the thrusters were quick in changing the direction of the ship but the blue streak still hit their vessel's habitation ring. There was a loud popping sound and the ship jolted roughly as some compartments depressurized. It was followed by a cacophony of bangs and scraping sounds of loose debris impacting the sides of the ship and within the compartment.

"Report!" Jack shouted into the intercom to see what the damage was. There was no response from anyone on the ship and Jack's heart sank deeper into his chest as he fought the urge to panic. Was everyone dead?

Then Garibaldi's voice crackled through the intercom, "The airlocks held. We're secure, I think."

"Increase velocity," Jack commanded.

"We're at maximum acceleration, Sir." Garibaldi reported, but to Jack it seemed as if nothing was happening fast enough.

Jack was intensively thinking about options how to best get away from the alien attack squadron, but the alien ships were tracking them. They were on a direct collision course with the alien lead ship, speeding towards it like one of the legendary kamikazes of World War II. Jack noticed that Zadak and Kitahari's shuttle was ahead of them. They must have fired their main thrusters as soon as Chan and Joe's shuttle was destroyed, and now they were considerably ahead of *Deep Explorer*. The smaller ship was certainly more maneuverable and didn't take as long to turn or accelerate. They zipped back and forth in front of *Deep Explorer* as if trying to draw fire away from her. The two ships roared towards the alien lead vessel at an incredible speed. The slightest error would cause an impact and destroy them all. Jack thought Lana was waiting too long to take evasive maneuvers and was about to say something when she vectored the ship to the left. For a second, Jack thought they were going to collide but the thrusters responded well. The shuttle spiraled off to the left just as another blue streak zipped by *Deep Explorer's* starboard side. Both vessels narrowly missed colliding with the alien ship.

The alien craft was slow to turn and follow them, giving them precious seconds and distance to escape, but the other four ships, the second wave, were approaching rapidly. One of them moved to the front and set a direct intercept course for *Deep Explorer*. Zadak and Kitahari were ahead with their shuttle and it seemed to Jack that they either wanted to collide with the alien vessel or move their ship between *Deep Explorer* and the alien vessel to protect *Deep Explorer* from enemy fire. Either way they were coming very close to the alien ship, but it appeared to ignore them. Apparently the aliens no longer considered the shuttle as a threat.

Lana tried to move *Deep Explorer* out of range of the alien ship but they were too close and needed more time to accelerate. She kept firing thrusters left, right, up, down to evade any weapon fire. Meanwhile Zadak's shuttle was extremely close, so close that it seemed that it would collide with the alien lead ship of the second wave any second. Suddenly, the shuttle swept to the side and ejected the ice clipper projectile, and then dove away from the alien fleet and *Deep Explorer* toward Titan. Jack watched anxiously: the projectile visibly jolted the alien ship but his hoped died when it didn't cause any obvious damage. Nevertheless, the jolt was perfectly timing as the alien ship fired at *Deep Explorer* an instant after the impact. The trajectory of the blue streak skewed to the side and barely missed them.

Lana fired upper thrusters, dove under the alien craft, and steered into empty space away from the planetoid, the Saturnian system, and the alien ships. Three of the four second-wave ships went into pursuit. The lead ship that had destroyed Chan's shuttle went after the other shuttle.

*Deep Explorer* continued to accelerate deeper into space and it became apparent that they were edging away from their pursuers. Another blue streak shot by their ship and then another. The second clipped the already damaged habitation ring. They weren't out of this situation yet. That much was clear. Two more bursts shot past them but with the increased distance and Lana's skilled piloting they were able to evade the attack.

Jack sighted in relief as the attack vessels pursuing them turned around and headed back to the alien planetoid.

"Vladimir, are you still on the imager?" Jack shouted through the intercom.

"Yes, sure!" he responded promptly but his voice was shaky and high-pitched.

"What happened to Zadak and Kitahari?"

"I lost visual and video contact when they moved behind Enceladus. The alien lead ship was still pursuing them and closing in."

## 5.3  Escape to Enceladus

Meanwhile, Zadak and Kitahari were on a steep descent toward the surface of Enceladus. They had sped away from *Deep Explorer* hoping to draw off some of the alien ships. It worked. One vessel had followed them, but now it seemed apparent that they would be dead soon. Or worse; captured alive! They had thought that Enceladus could give them an evasive advantage. The alien vessel in close pursuit had stopped firing at them, though. The shuttle was just too quick to be easily hit. For the aliens it was probably like trying to hit a dragonfly with a cannon. Kitahari continued to zigzag and dodge to prevent them from getting a clear shot. Getting hit wasn't their primary problem, however—their fuel reserves were dropping fast as they flew one evasive maneuver after another.

"There!" Zadak said to Kitahari. "See that ice geyser near the south pole? Go for it. It might be our only chance."

Kitahari steered the shuttle toward the geyser, which was nearly 1 km wide, shooting gas and liquids several kilometers above the surface.

"We are coming in too fast!" Zadak shouted. "Brake! Slow down! More reverse thrust!"

"I'm already on full reverse!" Kitahari screamed back.

The alien ship in pursuit had fallen back as they got closer to the surface but now it fired off another shot. The blue streak grazed the starboard hull with a sizzle and then impacted the surface of the moon, raising a blast of steam.

"Damn! Move through the geyser plume. We need cover."

Kitahari piloted the shuttle through the water plume ejected from the ice geyser. The shuttled rumbled and shook as it intersected the overheated water plume so violently that Zadak was worried the ship would break apart.

Kitahari kept the shuttle close to the plume and descended in a spiral path to the ground.

"Closer, closer to the source," Zadak ordered. "Only if we are close enough will our thermal signature be hidden by the heat of the geyser."

Kitahari struggled to keep the shuttle stable in the turbulence and uproar of the geyser. "We have to land. We're almost out of fuel."

"Do it!" Zadak agreed. "If we leave the geyser we're doomed."

Zadak turned off all transmissions and reduced power to absolute minimums as Kitahari lowered the ship closer to the surface. At that moment the shuttle was hit by a powerful water jet. The blast spun the ship around and it began tumbling downward. Kitahari worked feverishly to regain control and stabilize their descent. The surface was approaching fast and a crash seemed unavoidable. At the last moment Kitahari was able to straighten out the shuttle again, but they were still approaching the surface too fast.

"More reverse thruster!" Zadak shouted. "Damn!"

Zadak was interrupted as the shuttle impacted the surface so hard that the heat shielding exploded outward and part of the instrument panel shattered and came apart. Sparks shot all over the main cabin and loose objects ricocheted about the interior. Then the ship went completely dark.

Kitahari sat stunned and immobile as she struggled to focus on staying conscious. She felt abandoned and helpless—she was immersed in total darkness reminiscent of the witch's cage at the Complex. Only, here it was real. The pain of breathing brought her out of her daze. At first she thought her chest was crushed but soon she realized the crash had just knocked the wind out of her. She lifted up her head and felt a burning sensation in her face. She touched her face lightly and her hand came away wet from several cuts: only then did she become aware that blood was trickling into her eyes from a bad cut on her forehead.

Aside from cuts and bruises she seemed to be okay. "Zadak?" she asked as growing fear dominated her thoughts. There was no answer.

She fumbled for the button that would reset the emergency lights. After an initial flickering, the dim lights came on and revealed the full account of damage to the shuttle. Oh, *my God*, she thought as her eyes took in the scene around her. Zadak's seat had sheared off from the console and he had smashed into the control panel. Kitahari unbuckled herself and tried to stand, but a wave of dizziness sent her stumbling to the floor. She crawled to him on her hands and knees and unbuckled him as her blood dripped onto him and into her eyes, clouding her vision. She gently stretched him over the floor and studied his condition. She sighed heavily and fought back a sob of relief and happiness when she saw that he was breathing and alive, at least. He had a huge purple lump on his forehead, and various minor cuts, and his ankle was oddly bent and obviously broken. But he was alive.

She clapped her hands. "Zadak! Zadak, wake up!"

He groaned softly and tried to say something but the murmur was completely incoherent. He slowly regained consciousness and tried to sit up but fell immediately back. His eyes rolled back into his head and he fought to stay conscious.

"I think you have a bad concussion," Kitahari advised him.

"Yes, and my ankle is broken," he gasped as he grimaced in pain. "And my ribs too, I think. It hurts to breath. But you are not looking that good either. Are you all right?"

"Yes, just some cuts," she reassured him.

"What is our status?"

She looked over the instruments, did a major systems check and scanned the area. Blood was still trickling into her eyes so she ripped a piece of cloth off her uniform and tied it around her head.

"Okay, here we go. The main thrusters are damaged, but usable at 15%. We have hull integrity, but lost a lot of heat shielding on our starboard side. We may have enough fuel to reach orbit but we can't do much after that. The energy cells are down to 21%. Our oxygen tanks and carbon dioxide scrubbers are functional, but we don't have enough power to maintain air quality and keep the ship warm. Internal ship temperature is eighteen degrees Celsius and dropping. Sensors indicate we're sitting in a water-ice slurry. External temperature reads minus 130 degrees Celsius. It's going to get real cold in here very soon."

"What about the alien ship?" Zadak moaned.

Kitahari deployed the scanner that was barely extending out of the slush. "They are just sitting there right above the ice geyser. I guess they're waiting for us to come out, or they may think we are dead."

"We will be soon enough. The question is merely heat versus oxygen— whether we want to freeze to death or suffocate."

"Why do you have to be such a damn pessimist all the time?" Kitahari challenged him angrily. The tension had both addled her thinking and made her edgy. Zadak did not deign to reply. She took a deep breath and exhaled forcefully to calm herself. "I'm sorry," she said. "That was unnecessary. The situation has stressed me out and I'm not ready to admit defeat yet. Maybe Jack will come looking for us if they got away."

Zadak admitted that he could be pessimistic at times but he thought he was being realistic given their current predicament. He didn't want to tell Kitahari that she was being foolishly optimistic. Even if *Deep Explorer* got away, Jack wouldn't risk the ship by coming to look for them. Especially with them being pinned down here like two mice. If Jack had managed somehow to keep an eye on them during their flight all he would know is that they went around one side of Enceladus and never came out on the other side.

Zadak didn't wait for a response. He just allowed himself to sink back into unconsciousness.

Kitahari shivered as the rapidly-increasing chill began to affect her. She turned off all unnecessary power systems and then switched the power relays to turn off the air circulation system to warm up the ship, but she kept the setting low to conserve energy. She could keep toggling the environmental controls to give them either air or warmth but that wasted energy even faster.

She wasn't ready to give up yet. If the alien vessel above left, she was sure *Deep Explorer* would come looking for them. She figured they had about twenty-five minutes of usable air before she had to switch back to air control. But if she could figure out a way to manually dispense the oxygen then maybe she could buy them more time and energy. And if Jack came back they would need to have a functioning ship. Therefore, she would try to repair what she could in the time given. A plan of action was now set in her mind. She moved about in an effort to make her hopes a reality and began humming a few lines from "We Three Kings" by John H. Hopkins:

*"Myrrh is mine, its bitter perfume*
*Breathes a life of gathering gloom*
*Sorrowing, sighing, bleeding, dying*
*Sealed in a stone-cold tomb"*

*****

## 5.4 Death and Defeat

Jack's full attention was on his own ship.

"Garibaldi, turn off propulsion. They won't be able to catch us now and we need to conserve fuel. You and Pablo assess the damage. Lana, plot a course that takes us on a wide trajectory back to the Saturnian system, but keep us well out of reach of the alien ships."

"Jack!" Carmen yelled through the intercom. "You should look at the video coverage from inside Chan's shuttle. Hurry! Something is happening there."

Jack flipped through the controls of the monitoring screen to access the ship to ship video link. It was hard for him to see any details at first. The video camera showed little sparks and flashes of light, but there were utilities and tools floating in the main room of the shuttle and they were partially blocking the view. He couldn't really make out much. At some point Jack could see a body, which looked like Joe's. But, yes, something was happening there. The debris cleared a bit and Jack saw that something was melting through the hull of the battered shuttle to create a cavernous portal. The sparking stopped and from the hole something emerged.

Jack watched, mesmerized, as three aliens in spacesuits entered the craft and began moving things out of the way. Due to the spacesuits and obstructed view, he couldn't see much but he recognized that they were quite a bit larger than humans and had fours arms and two legs, or six legs, depending on how one would interpret their physiognomy. They moved about on either two legs, or on four, or on all six. Not much else was visible other than that there was an apparent division of their body into big sections, like an abdomen and torso, and they had a distinct head. They moved rather slowly compared to a human. Their whitish space suits concealed any further details. Jack could now see Chan's body, also. The aliens moved the human bodies next to each other and used some instruments on them, apparently scanning. "They are studying us," Jack thought.

He continued to watch as the aliens crammed Joe and Chan into a single large bag-like container and then exited the shuttle. Jack could only imagine what they would do with the bodies. They were probably destined for some alien laboratory for dissection and analysis. Or maybe they would be eaten as if they were nothing more than animals. Maybe they would be dragged through the halls of the alien vessel and mutilated in some sort of victory ritual. Jack tried to put such thoughts out of his head. At least Joe and Chan were dead and didn't have to endure whatever horrors the aliens intended for them.

Jack called Vladimir on the intercom. "Has this all been transmitted to Earth?"

Vladimir's pale face appeared on the screen. He just nodded.

"Garibaldi and Pablo, I need that damage report."

Then, Jack sat back. They were safe for the moment, he reflected. But their situation was still very serious. They had limited supplies and a damaged vessel. They had burned a lot of fuel in the encounter. The alien invasion force was between them and Earth. And yes, what about Earth?

This was a serious threat to humankind and possibly the whole planet. The aliens weren't friendly and apparently couldn't be reasoned with. It was regrettable that mankind's first encounter with intelligent life forced them into war. But why, Jack asked himself, why can't there be a peaceful coexistence? Why must there be war? This path can only lead to the annihilation of one race or the other.

Jack tried to lay out in his mind mankind's tactical situation. The alien ships were much more powerful and numerous than anything Earth had. The only edge his ship had was that it was faster. That may have helped them now, Jack contemplated, but it would be of little use if you want to defend a stationary target such as Earth. Earth wasn't completely defenseless though. There was the fleet of ships being built back home, but they were too few and no one knew if laser weapons would even have an effect on the alien vessels. There were the tactical nukes and intercontinental ballistic missiles, too. Jack was sure they could quickly be adapted to space combat. They question is whether or not they would be effective.

Jack wasn't sure if he could still play a part in defending Earth or not. The only thing they could do right now was to monitor each step the aliens took and try to learn about their possible weaknesses. Perhaps the aliens would stay here and leave Earth in peace, but he considered that unlikely.

"Vladimir, anything new from the planetoid and the alien ships?" Jack asked.

"Their main force is just about to reach the moons of Saturn. The planetoid has slowed down."

Perhaps there is still some hope that the aliens won't continue to Earth, Jack thought.

Jack finally contacted Earth HQ. The mood was somber, since the whole encounter had been closely monitored from Earth.

"What is the condition of your spacecraft?" Colonel Haggerty asked.

"We are assessing the damage," Jack responded. "All primary systems seem to be fully functioning and the damage seems limited to the habitation ring. We have lost contact with Zadak's shuttle and feel it is likely that they were destroyed. Do you have any other information?"

"No, we lost contact a while ago. No video transmissions anymore. The last info that we received was that they tried to hide somewhere on Enceladus. Then the transmission went off. Either they were destroyed, or the equipment was damaged or malfunctioned. But we don't have any confirmation either way."

Haggerty paused for a moment to let the info settle into Jack's mind and then he continued. "Don't worry about that too much now. Your higher priority is to save yourselves. You provided us with invaluable insights and all of us here on Earth appreciate your bravery and sacrifice, but I have one issue I'm not satisfied about. Why didn't you use your laser?"

"There wasn't enough time, Colonel. We were struggling not to be hit and didn't have the opportunity. To do so would almost certainly have resulted in our destruction even if we managed to use it successfully."

"I understand. However, we absolutely MUST know whether the laser is any good in combat against them, especially if the aliens should come toward Earth. You have to find out! I also understand what I'm asking you and your crew to do."

He was asking Jack and his crew to die in a blaze of glory—that was what he was asking them to do. Damn! Whether Jack liked it or not, the Colonel was right and there wasn't any other path. Earth had to know if the laser was effective. "I'm sure we will have a chance to test the laser soon. Over and out."

Jack reflected on the last hour. Everything had gone so fast, yet it appeared that the time before the encounter was a lifetime ago. He wished he could get to Enceladus to at least search for Kitahari and Zadak, but that seemed impossible. He was stranded with his crew further away from Earth than ever before. His crew would look to him for leadership, but what did he have to offer? The promise that they would die bravely in a meaningful suicide mission and that they would be remembered forever as heroes. That didn't sound very inspiring.

He was suddenly interrupted. "Jack, please come down. You've got to see this!" Vladimir shouted excitedly through the intercom.

When Jack arrived, all the remaining *Deep Explorer* crew was there, too, except Lana, who remained in the cockpit.

"Look!" Vladimir said. "The alien planetoid is inserting itself into orbit around Saturn, exactly where Tethys is located, and on a direct collision course for it!"

They all stared at the images. As the KBO approached Tethys several huge purple streaks emerged from the planetoid and Tethys went careening out into space as it was pushed out of orbit. The crew was stunned as they watched Tethys move away from Saturn into deep space. The moon would spend the rest of its existence as a runaway planet. The power it took to move a mass the size of Tethys was beyond the collective imaginations of all of them.

"The alien planetoid is taking Tethys' position. With its diameter of about 1000 km it is nearly the same size as Tethys and thus will have a stable orbit," Carmen commented through her shock.

Garibaldi moved the imager toward Titan. "What the heck are the alien ships doing there?"

Three of the ships were moving into a triangular formation inside the upper atmosphere and initiated a pulsating whitish streak.

"They are targeting the Void!" Vladimir shouted.

"Analyze the pattern. The pulse!" Jack blurted out in a hurry and they rushed to turn on all scanners and recorders and aimed them toward Titan.

The pulsating pattern continued for about twenty seconds, then appeared to get more violent until there was on implosion on the surface producing a huge flux of subatomic particles.

"I think they just blew up the Void," Vladimir said.

There was a long silence in the room. Finally, Jack instructed Vladimir to transmit all sensor recordings to Earth. After they had compiled and sent their readings, Jack asked Pablo about the condition of the habitation ring.

"The fitness room and three of the habitat cabins are destroyed; yours, Lana's and Joe's, but otherwise the damage is relatively minor. The airlocks are secure and holding, but we won't have access to the full habitation ring until we can get it all fixed and pressurized. The alien weapon appears to be some form of plasma energy. It passed right through the habitation ring like a hot knife through butter and left a soccer-ball sized hole. It also damaged the ventilation system but I shouldn't have a problem bypassing the damage. Most personal supplies in the affected rooms appear to have been sucked into space."

"That's good news then for the most part," Jack said. "Try to patch the damage as soon as you can. We don't want to be all bunking in the conference room. Everyone else attend to your duties. We'll mourn the dead later." Jack excused himself to return to the cockpit with Lana. It wasn't long before he got a frantic call from Carmen to come to the lab: she was completely upset.

He entered the bio-lab to find her and Vladimir intently studying the Titan slugs. "After we were hit and out of immediate danger I checked on the slugs. They were all right. Now, look!" She went with the scanner over what appeared to be their remains. "Only slush, organic slush, all dead."

"It must have to do with the Void," Vladimir noted. "That is the only logical explanation. The aliens destroyed the Void and without the Void, all Titan life turned into slush and decomposed. Even your slugs that were not on the surface of Titan. The Void must be intrinsically connected to life, just as Zadak proposed earlier. The bastards destroyed all life on Titan, Jack!"

"Why?" Carmen asked, in tears.

"I guess they don't want any competition. Or, who knows, maybe it's their food source," Vladimir hypothesized.

"If true, this is extremely disturbing," Jack added.

Garibaldi's voice popped up on the intercom and interrupted them during their analysis. "I think you should return to the imager. Something is going on out there."

Everyone returned to the imaging room. Through the imager they saw that the aliens were massing ships near their home world that had replaced Tethys.

"Those are not meant for us, or what do you think?" Garibaldi asked.

"I don't think so," Jack responded. "They must have other intentions."

They watched for several minutes as the ships continued to maneuver into a larger fleet. When the fleet of ships started moving in the direction of the inner Solar System, the alien's intentions were clear.

*First they destroyed all life on Titan, and Earth is next*, Jack thought. He studied the map and their trajectory. Perhaps there was some way to push through the alien corridor and move toward Earth. Not that they would be of much help, but it felt like the right thing to do. Then Jack noticed that there was only one alien ship near Enceladus in a geostationary orbit over the south polar region.

"Vladimir!" he said, getting excited, "Can you confirm that there is only one alien ship near Enceladus?"

Vladimir hastened over to double-check the imager. "Yes, that's correct. I don't think they consider us a threat anymore. And it is definitely not Zadak's shuttle! They are pretty much ignoring us."

"I have an idea. If we just alter our trajectory a little bit we could swing by Enceladus and close to Titan as well. What do you think?"

"Yes, but what do you have in mind?"

"I think its time to test our laser and look for our missing friends at the same time."

"You mean mosquito tactics? Sting and run?"

"Hopefully more than that. But it is critical that there is only one ship over there, otherwise it will be suicide. This may be our best chance to test the laser and we might not get another one."

"I will continue monitoring and let you know if the situation changes. I will also continue trying to get into radio contact with Zadak and Kitahari."

"I think we have a plan!" Jack said with some excitement, the first excitement since the encounter that had gone so badly. "I'll contact Earth."

## 5.5 Counterstrike

Jack reached Colonel Haggerty and Dahai on the telecom link and reported his observations in detail. "Well," the Colonel said, "if they are coming, we have to have everything ready in six and a half months."

"You have to protect the Void at all costs," Jack emphasized.

"Well, that's not for certain."

"We will verify. We have a plan to attack the single alien ship that is hovering over Enceladus, make a search for Zadak and Kitahari, and then swing by Titan to confirm the destruction of the Void and Titan's life."

"That's a good plan," the Colonel agreed. "We really need to know about the effectiveness of the laser."

Jack replied, "Knowing that will be truly helpful to us all. If we don't do anything here, we will be dead sooner or later anyway. This way we will still be of use. Even if the laser does not work, you need to know!"

"Good luck, Jack, and keep us informed."

Dahai nodded seriously but with a slight glimmer of a smile as they logged off.

Jack called everyone to meet in the briefing room. *Deep Explorer* was only a few hours from Enceladus and they had to be ready for battle when they arrived. Jack wanted to hear everyone's thoughts on a battle strategy. Vladimir suggested having the ship wobble a bit so it would appear that they were not in full control of the ship, suggesting battle-damage. Jack agreed to do anything needed to allow the aliens to underestimate them. Hopefully the aliens would not consider them a threat and move ships for an interception. Garibaldi was a bit concerned about the fuel situation for the major propulsion system, but Jack insisted on using any resource available for the planned maneuvers—better to use fuel wisely when NOT being shot at, than to take too few preliminary evasions and wind up using it while being used for target practice by the aliens!

Time passed quickly as they approached Enceladus. They were as ready as possible given the situation. As they moved toward their intended target, Vladimir reported that no alien ships were moving towards them. Even the lone alien ship hovering in orbit over Enceladus appeared to just be waiting for their arrival. When they were fifteen minutes from the alien ship, Garibaldi charged up the laser. Pablo initiated a random sequence for firing the left and right thruster, partially as evasive maneuvering, partially intended to appear damaged and halfway out of control.

"Laser ready for firing!" Garibaldi announced through the intercom about ten minutes before interception.

"We hold." Jack said. "We will fire the laser one minute before interception, not a single second earlier. We need the full power."

Two minutes until interception. The seconds ticked by like little eternities. Then suddenly the alien ship fired a blue streak. Due to the random wobbling and Lana's quick reaction, the ship was only hit at the right back thruster, but it caused an explosion and shrapnel was flying around in the engine room; one chunk hit Garibaldi in the upper leg.

"Man down!" Pablo screamed.

"Hold your position!" Jack shouted back, "5, 4, 3, 2, 1, fire laser!"

Pablo pressed the release button and the laser fired. The beam hit the egg shaped ship right in the front center. Jack swallowed. He couldn't see any visible effect. "Continue firing, and go to full speed," he shouted.

After a long few seconds, the hull of the alien ship simply melted, and the ship opened like an overripe fruit. They could see material from inside being sucked into space.

"Kill the laser!" Jack commanded. "Initiate reverse propulsion. We stop and board that ship. The aliens had to use space-suits just like us, so they are most likely all dead in there. Vladimir, suit up and get ready. You and I will go over. Carmen, stay on the imager and monitor for any intercepting alien ships. You are in command."

"This was not planned!" Carmen protested. "We are sitting ducks and this is much too dangerous."

"We won't get this chance again. Any information we can get about the aliens will help Earth. Continue the automatic transmission to Earth! Send them everything! And make doubly certain that they understand that the laser worked perfectly! It's a great weapon!"

Pablo yanked the bit of metal from Garibaldi's leg and then put on a bandage.

Lana moved *Deep Explorer* skillfully to sit right in front of the gaping hole through the hull of the alien ship. The alien ship looked completely disabled. Then Vladimir and Jack climbed into their space suits, left the ship and propelled themselves with rocket backpacks into the hole. The bizarre layout of the ship confounded them and the scattering of debris mesmerized them.

"You have a maximum of ten minutes. Then we have to get out of here," Carmen said through the telecom. "We've drawn the attention of the alien ships and five of them have started moving towards us."

Just at this moment, Carmen received another transmission. "Shuttle 2 here," Kitahari's excited voice crackled over the comlink. "Boy, are we glad to see you guys."

"Kitahari?!" Carmen screamed in joy. "It's great to see you, too. How are you?"

"We have heavy damage and Zadak is injured, but I will try to get the shuttle up into orbit. That's about all it can do now."

"Can we help you in anyway?"

"I don't think so. Just stay where you are and I'll try to steer in very slowly."

"Not too slowly though, Kitahari. We've got five alien ships approaching fast and we need to get out of here as soon as possible!"

Meanwhile Jack and Vladimir wandered through the alien ship. They stumbled their way amidst large metal parts and pieces of some strange soft alloy until they found an alien body that was stuck in debris. The insectoid alien body was white, almost translucent. It seemed fragile with the exception of the head, which had a large jaw that was ringed by a pair of pincers. It didn't appear to have any ear or nose equivalents, but it had small eyes and a bundle of ten antennae on each side of the head. The body consisted of a thorax and abdomen, both amazingly soft. The six legs originated from the thorax and were of hard organic chitin-like material. The dead body had a translucent liquid spilled over it and around it, which Vladimir analyzed with his scanner, while Jack continued to look around. He found a few more bodies that apparently had all suffocated in the vacuum. Some of them looked quite different from the one Vladimir analyzed, reminding him of caste differences in Earth's bees and ants. The body proportions were different between individuals. One had an abdomen at least triple the size of the other bodies and a larger head. Also, some of the bodies did not have pincers. Jack walked back to Vladimir.

"What is the liquid?" he asked.

"Organic compounds mixed with ammonia and water."

"We have to leave here in a few minutes!" Jack said.

"There is no way I can get a good idea of the physiology and biochemistry of that guy in a few minutes." Vladimir countered

"Then we have to take one with us."

"Jack, are you serious? This thing is enormous."

"There's no time to argue unless you have a better idea."

Vladimir just grunted and murmured something in Russian beneath his beard as he grabbed the creature by a hind leg and started dragging it. It reminded him of what the aliens did to Chan and Joe. We are no better than they are, he thought. Except that they fired first.

"Jack, we have to leave in three minutes," Carmen insisted nervously. "Don't make me leave you behind."

Jack couldn't tell if she was joking or not, but decided that she probably wasn't. The safety of the ship was more important than his and Vladimir's lives. "We're coming as fast as we can."

"We are hooked up, Carmen," Kitahari said through the comlink. "But I need some help. Zadak got hurt when we crash-landed."

"Pablo, get down there fast and help her!" Carmen ordered. "Garibaldi, recharge the laser."

"Will do," Garibaldi acknowledged. "But it will take at least thirty minutes."

"Forget the laser then."

"Lana, as soon as we have Jack and Vladimir, move us behind the damaged alien vessel. We'll need some cover."

"Jack, you need to get out of there, now!"

"We are on our way", he responded and Carmen saw the two men with the alien body climbing out of the hole a second later. It would still take them another minute to get on board *Deep Explorer*. And they didn't have a full minute. The alien vessels were closing too fast.

The alien vessels aligned themselves again in their usual formation with one ship in the front and four in the second wave.

"We're in!" Jack yelled through the comlink and began shedding his spacesuit. A split second later the propulsion system ignited and boosted the ship away from the approaching aliens. Lana steered *Deep Explorer* towards Enceladus and used the moon to give their acceleration a gravity boost as she tried to put the moon's 1000 km of rock between them and their pursuers.

Jack hurried to the cockpit to assist Carmen and Lana. The five alien ships were still gaining fast, and *Deep Explorer* was still accelerating. They got within firing range and were still closing but did not fire. They were going to wait until they got as close as possible, Jack thought. Damn it! He cursed. If they could only have gotten out thirty seconds sooner. The aliens still held their fire. "We've matched velocity with the alien vessels," Lana reported to Jack.

Yes, Jack reveled. And now they would continue to accelerate and outrun their adversaries.

## 5.6  Darkest Hour

They were closing in on Titan. Vladimir used the scanners to monitor Titan during the close flyby. Carmen had instructed Lana to pass as close as possible, but not to waste time with anything else. This has to be good enough, she thought, to assess whether there is still life on Titan. Jack monitored the five alien ships that were on their tail. They had just passed Titan when from their hiding place behind Titan three alien ships ambushed them and sent multiple blue streaks in their direction.

"Damn it!" Jack squawked when he saw the alien vessels firing. "Evasive maneuvers."

Lana jerked the controls and the ship lurched sideways. She was able to evade the first two streaks, but the third one hit the propulsion system and took out the starboard thrusters. Fire erupted into the fuselage, and several sheets of hull plating blew out. Zadak, Kitahari, and Pablo, rushed to extinguish the fire, but a pocket of the remaining fuel ignited and resulted in an explosion that sent the ship into a spin. Shrapnel went screaming through the air in all directions. Pablo and Kitahari were partially protected behind a bulkhead, but Zadak, slowed down by his previous injuries, took the full blast. His body hurtled across the room and smashed into the airlock door.

Meanwhile, Lana fought to restabilize the ship and maintain their trajectory.

"We need to go below and see if we can help," Jack said. "Just get us out of here," he said to Lana, and sprinted with Carmen out of the cockpit to the engineering section.

Lana regained control of the ship and pointed it away from the pursuers. She never saw the fourth blue streak fired from the alien ships as it sped toward *Deep Explorer*.

The plasma bolt went through the cockpit as easily as a rocket would drill through a piece of cheese. She never even felt the surge as it blew through her body and she didn't have to face the painful death of sudden vacuum as the air exploded out of the cockpit. Death came instantaneously.

And *Deep Explorer* continued to follow its trajectory, heedless in the absence of its pilot. Thanks to its momentum, it was soon out of reach of the alien ships.

Jack and Carmen met Vladimir just outside the engineering compartment and as they rushed inside they were greeted by a blast of hot air. Choking smoke filled the room and stung their eyes and throat. Kitahari and Pablo fought the fire as they were wracked by spasms of coughing. Jack and Vladimir jumped into the thick of it to try to knock the fire down. For a while Jack was worried the fire would spread, but slowly the combined efforts of all five of them managed to suppress it.

When it was over Kitahari trotted to where Zadak was laying, still next to the airlock, and the others followed. Kitahari knelt down beside him and cradled his head. Zadak was bleeding from various wounds and his face was scorched with bits of dried and blackened skin peeling away. But the fatal wound was the square tubing imbedded in his chest. Zadak opened his eyes and saw Jack.

"I guess today wasn't my lucky day after all. No one told me that we have to fight a war out here," he said with a faint smile. Kitahari tried to quiet him, but he continued with a weak raspy voice. It was obvious that blood was filling his lungs, but there was nothing anyone could do.

"Jack, you need to figure out what the Voids are all about. Only this way can you save Earth." He tried to speak again but his words died away in an incoherent gurgle as frothy blood oozed out of his mouth. Then his eyes closed slowly and Kitahari began sobbing quietly. All they could do was watch as he died in her arms.

Meanwhile, Garibaldi was able to reduce the temperature in the impacted thruster area to prevent another explosion or fire. Satisfied that the danger was minimized he began a ship wide diagnostic systems check. That's when he noticed that there was no atmosphere in the cockpit.

"Lana, come in! Lana, come in!" There was no response. Garibaldi's face reddened. He ran toward the cockpit and grabbed a spacesuit enroute. Carmen ran after him, but even though Garibaldi had a bandaged leg, she had trouble to keep up. When they reached the airlock door to the cockpit, Carmen helped him put on the suit and then left the hallway and got behind another airlock. It took Garibaldi a few minutes to override the safety protocols and evacuate the compartment behind the cockpit before he could enter.

The scene that greeted him threatened to overwhelm the tentative hold he had on his emotions. Everything not permanently affixed was completely destroyed or had floated off into space. Two soccer ball size holes marked the entry and outlet points of the alien weapon. Lana was lifeless, still strapped into the pilot's chair. Garibaldi moved to the other side of Lana and saw a blackened hole where her lower abdomen used to be. Only part of her spine and some other tissue remained to hold her together. Her face was swollen and deformed from sudden decompression and Garibaldi fought against the urge to cry when he grabbed her lifeless body. He unstrapped her from the chair. Her body was already cooling and stiffening.

Garibaldi picked her gently up and carried her into the compartment behind the cockpit. He shut the airlock door behind him and had to wait a few minutes for re-pressurization. Carmen rushed into the room to help Garibaldi take his helmet and suit off. There was a smell of burned flesh in the air. Garibaldi just left the suit on the floor and carried Lana out in his arms. Jack and Vladimir were already waiting in front of the door and they wordlessly followed Garibaldi to the conference room, where he laid her body next to Zadak's.

Jack tried not to think of the deaths and loss. He tried to tell himself that their sacrifice had meaning and a purpose. It was true, he knew, but that truth seemed so hollow, bitter and pointless. They would be safe now for a while, though. They had done what they had to do and now they could move on.

The tears came then, not just Jack's. Garibaldi, Carmen, Pablo, Vladimir and Kitahari sobbed out their pain as their emotions overwhelmed them. Now that they were out of danger they allowed themselves to face their losses.

They silently mourned their dead.

"What are we doing now?" Pablo finally asked.

Jack, who was holding Carmen, responded. "We need to inform headquarters of everything that has happened here. They already received the video feed of our encounter with the alien vessel, but I will report the details of our combat and on the laser weapon and its effect, and on the current status of our vessel. We will have to deal with our departed friends soon enough, but we can't afford to sacrifice our other objectives. Garibaldi and Pablo, you have to work to get the ship fixed, the most important systems first. Vladimir, you need to focus on your scan recordings from the Titan flyby and confirm that the aliens destroyed the Void and all life on Titan. Carmen, you will need to start analyzing the alien that we brought from their ship. Have Vladimir help you getting it to the lab. Kitahari, you can assist her. I will see to Lana and Zadak."

"What do you intend to do with them?" Kitahari asked.

"We will have a memorial service later. First, we have to take care of our vessel. Otherwise there will be no one there for a memorial service. Keep me updated on your progress."

After three hours they came together for a quick briefing. Garibaldi and the other crew members were able to seal the holes in the cockpit and re-pressurize it. It was not yet functional, though, and needed much more work. However, Pablo figured out a way to reroute thruster control to the engineering compartment so they could steer. They were not able to lay in an exact course, but could probably be successful within a nautical degree or two. It would be enough for the moment.

Vladimir was able to confirm that there was no trace of the former Void on Titan and that life probably did not exist on the moon anymore. His spectral scanners revealed the same dead organic signature all over the moon that matched that of the dead slugs in the lab. He also confirmed that the alien planetoid, which he now referred to as "Tethys II", was in a stable orbit around Saturn. Pablo was also able to get access to the habitation ring so that everyone could get ready for the memorial service. Only Jack's room was totally destroyed and it would take some time before he could use it again. Jack cleaned up the bodies and dressed them in clean clothing. He asked for everyone to clean up as best they could and get ready for the memorial service in an hour. Then he sent their current status and findings to HQ.

The hour of departure for their fallen companions had come. Jack, Carmen, Garibaldi, Pablo, Valdimir, and Kitahari assembled in the cargo bay and stood in a vigilant semicircle around Zadak and Lana.

Jack felt it was his solemn duty to speak on behalf of the crew.

*"We come together to say goodbye to our dear friends Zadak Szodan, Lana El'Barais, Jeau-chin Chan, and Joe Palati. They answered a call beyond duty and gave more than was expected or required, and gave it with little thought to*

*worldly reward. Today, the frontier is space and the boundaries of human knowledge. Sometimes, when we reach for the stars, we fall short. But we must pick ourselves up again and press on despite the pain. Fortunately, there are still many with courage, character and fortitude that will carry on the journey begun by our friends."*

Jack looked around at his comrades and then continued:

*"We remember Zadak Szoda, born and raised in a small Serbian village, throwing rocks at military personnel when he was just seven years old." Jack smiled for a moment. "Zadak did not like authority, but he was a great scientist and never ran out of ideas and inspired the excitement of all his fellow scientists.*

*We remember Lana El'Barais who dreamed already as a kid to be a pilot, somewhere in the slums of Lagos. But her dreams and dedication took off, and her wishes were fulfilled by piloting the largest starship of our time.*

*We remember Jeau-chin Chan, always dedicated to heal his follow trainees and crewmates whether physical or psychological wounds. Loyal friend, without you we wouldn't have gotten as far as we did.*

*We remember Joe Palati. His humor made us always laugh even in the darkest hour. He never shied away from any risk and was willing to pilot the unarmed shuttle into the alien's wrath.*

*Today, we promise that their dream will live on, and that we boldly will face whatever may be out there. May their spirit guide us. May their spirit be united with the spirit of the universe. Their sacrifice was not in vain, but will help us and our brothers and sisters on Earth to prevail and preserve our species. We will make you this promise. We will do anything in our ability to protect this for what you gave your life. We will. We have to prevail. You can rest in peace."*

Then Jack nodded to Kitahari, and she passed out copies of a flyer containing a hymn. She started singing the refrain first on her own:

*Shepherd me O God, beyond my wants,*
*beyond my fears, from death into life,*

and then they prayed together:

*"God, is my shepherd, so nothing shall I want,*
*I rest in the meadows of faithfulness and love,*
*I walk by the quiet waters of peace.*
*Gently you raise me and heal my weary soul,*
*you lead me by pathways of righteousness and truth,*
*my sprit shall sing the music of your name.*
*Though I should wander the valley of death,*
*I fear no evil, for you are at my side,*

*your rod and your staff, my comfort and my hope.*
*You have set me a banquet of love in the face of hatred,*
*crowning me with love beyond my power to hold.*
*Surely your kindness and mercy follow me all the days of my*
*life, I will dwell in the house of my God for ever-more."*

They closed by singing *Shepherd me O God* together and then released their comrades to space.

Jack had to admit that the memorial service affected them all very deeply. But of course it had to: they were all close to their shipmates. They all cared deeply for each other. Jack hoped that would help him and his crew to get over the loss of their colleagues and friends.

Then they left silently to their quarters, except for Jack. He went to the inoperable cockpit since he couldn't return to his room and positioned himself in meditation posture when someone knocked.

"Yes."

Carmen opened the door and sat down next to him. "Where do we go from here?" she asked softly.

"We will meet tomorrow morning and discuss that."

"I don't know whether I can. There is too much sadness in me. It's just so overwhelming!"

"You can. You are stronger than you know, Carmen."

Jack moved over next to her, and gave her an intense hug. Then he wiped away the tears on the ends of her eyelashes.

"We are still alive, you know. And we have to celebrate life. We cannot succumb to despair and depression."

"I feel so alone in space, so far away from Earth."

"You are not alone, Carmen. We have each other."

She rose then and took him by the hand. "Come with me, Jack," she instructed.

Jack rose and followed Carmen to her cabin. She turned off the lights and they tenderly allowed their bodies and their spirits to move closer and closer together. And finally unite.

# 6  Tethys II

## 6.1  Phoenix

They all met in the conference room the next morning. Jack evaluated his remaining crew members as they waited for the meeting to begin. Kitahari and Vladimir stared numbly at their shoes. Carmen rested her head in her hands and stared blankly at Jack. Garibaldi waited quietly to give the engineering status report. His face was grave and haggard, with sunken, red-rimmed eyes. The strain of battle and the loss of their companions showed clearly on everyone's faces. They were exhausted. And it seemed to Jack that the pall of defeat and loss was dragging them all down into hopelessness. But it wasn't over yet and Jack couldn't allow that to happen.

"Garibaldi?" Jack prompted the engineer to begin his report.

"The vessel is severely damaged. The starboard thruster is permanently disabled. We can't fix it without spare parts from Earth. The cockpit needs to be rewired and partly rebuilt, but I think we can get 86% functional recovery on the forward systems—but it will take another three weeks and we'll have to scavenge a few parts from other nonessential parts of the ship. We can have the back port thruster operative in about the same time frame, but I don't think we will be able to push it beyond 30%."

"What about fuel reserves?" Jack asked.

"We are pretty much out. We have enough fuel for course corrections and perhaps one tenth of an ignition burn for acceleration, but that's it."

"So, there is no way to stop for repairs or anything?" Kitahari interjected.

"No, we have to continue at this velocity and we have to decide which way to go, back to Earth or …" Garibaldi did not finish. He just shrugged his shoulders tiredly.

There would be no point in encountering the aliens again, Jack thought. They had done their duty for king and country. For now they would monitor the aliens' activities until they could repair the ship as well as they could. If they could get back to Earth and get the ship fully repaired, then the vessel might be of some tactical use, but not now.

"How about the shuttle?" Jack took back the initiative.

"We can fix that as well. But it will take a while, probably another two weeks on top of everything else before we can have it all fixed up. Or we could just scrap it for spare parts and transfer the remaining fuel back to *Deep Explorer*."

"We'll try to fix it for now, but only after all the other repairs are done. Okay, we have a lot of work to do, people."

The assembled group remained silent.

"Carmen is in charge of analyzing the alien body, trying to understand their physiology and mental capabilities. Vladimir will continue to investigate how the aliens destroyed the Void on Titan. Vladimir, you and I will also analyze strategies for future encounters. If not for us, then for the folks back on Earth. Everyone else's first priority will be repairing the ship. Let's get going!"

The next days and weeks were filled with the reconstruction. This required multiple space walks. Garibaldi ran himself ragged as he supervised this work, with Pablo and Kitahari, and sometimes Jack, to help him out. They worked around the clock. After having fixed major parts on *Deep Explorer*, they moved on to repair the shuttle. Pablo laid in a course that kept their vessel in the Saturnian system but far enough from Tethys II to stay out of the reach of the aliens. Earth HQ agreed with them continuing to monitor the aliens, while repairing the ship. They delayed any decision on how to proceed until after the repairs. Jack discussed various options with headquarters and his crew. The primary options were either to fly *Deep Explorer* back to Earth or to remain in the Saturnian system and continue taking observations.

The mood was somber though everyone tried to appear upbeat. Carmen spent long hours in the laboratory analyzing the alien. Their blood equivalent was a salty liquid containing about fifty percent ammonia and 50 % water. The enzymes she could separate from their body chemistry had the highest catalytic activity at temperatures of minus forty degree Celsius and two and a half bar. She concluded that this had to be close to their usual living conditions.

There were no sun protection pigments that Carmen could find and she hypothesized that they evolved to live either deep in the subsurface in the absence of sunlight or on a world where sunlight never reached the ground. Or maybe they were strictly nocturnal. She speculated that this might also be the reason why they were so sensitive to focused laser light in the visible range. Their antennae were the most sensitive organs and contained thick cords of spidery nerve clusters. The brain was small, too small, Carmen felt, for an advanced species. Based on their physiology, bodily movement agility, and brain processing capability, they should be rather slow; one-third to one-fifth of the processing and movement speed of humans.

"How come they are beating us then?" Garibaldi inquired.

"Simply because they have been around for much longer than us and have had a lot of time to come up with more advanced technology. Just because they are physically slow does not mean they are dumb," she responded.

"But their brain is small."

"Yes, but they seem to make up for that by superior communication. Their communication center within their brain is nearly half of their total brain, more than three times larger than in a human being. Besides, Jack and Vladimir noticed various types of aliens, when they were on the alien ship. Perhaps they use some kind of advanced separation of tasks, some of them are evolved or genetically engineered to do physical activities, others warring activities, others thinking. That seems to be a good assumption but with only having parts of one type of organism, that is impossible to tell. The one I have has so much strength in its legs that it can easily lift up ten times its own bodyweight. Can you do that? The worst mistake we could make would be to underestimate them."

"I don't think anyone will make that mistake after our experience," Garibaldi added.

Vladimir was the one most often in communication with Earth with the exception of Jack. They were successful in identifying the energy pattern of the alien craft that destroyed the Titan Void. This was a hopeful and depressing discovery at the same time. Hopeful, because they knew now how to destroy a Void; depressive because they had no means to produce such an energetic pulsating pattern. There was certainly no way they would be able to generate such energy on *Deep Explorer* and launch it against Tethys II, even if the alien Void were located on the surface, which it wasn't. Vladimir realized also that the so-called "death frequency" could very well be different for different Voids. It felt to Vladimir that it was a race against time. Perhaps they were smarter and faster than the alien species, at least in principal, but how could they make up for thousands, perhaps millions of years of evolution and advanced technology development in a matter of months?

Jack provided headquarters with frequent updates on their status and Colonel Haggerty, Dahai, and Mike Hang assured him that Earth was making feverish preparations to be ready for the alien invasion fleet. By now, the invasion fleet of fifty-one vessels was nearing Jupiter and its satellites.

They clearly are adapted to cold conditions, Jack thought, so they can't live on Earth. They only come to destroy any future competition. There must be a way to beat them. The situation was depressing, and frustrating at the same time. It made Jack afraid of the possibility that the human race faced extinction. And his fear made him angry at the same time. We won't go quietly into the night!! But what could he do? The fate of every living thing on Earth was at stake.

He occupied himself mostly with monitoring the progress of the crew repairing the ship and helped out wherever he could. The most common problem was finding parts: they had to improvise a lot. Too often they would have to cannibalize the working parts they needed from functional parts of the ship. Thanks to Garibaldi's ingenuity they found solutions to most problems. If it wasn't for the looming uncertainty and alien threat, this time would have been one of the happiest of Jack's life. He treasured the closeness he had with Carmen and they were open about their relationship with the other crew members. The crew seemed to be happy that Jack and Carmen had found one another, and in some strange way they were relieved that in such dark times it was still possible to carve out some light and happiness. Once, when they were lying together, Jack confided to Carmen his admiration about the human spirit and his crew, and just hoped that they would not witness the end of human civilization in the next few months.

On another occasion Carmen asked Jack, "Do you think there is a God? If there is, wouldn't he—or she—or it—help us?"

"I don't know," he responded. "I've done quite a bit of thinking about that since I have been at the Complex. I believe so. There are so many things you can't explain without the existence of a higher being. I imagine God as a union of all living beings. Look at the cells of your body," and he pointed at her body. "Zillions of them. Each of those cells is individually alive, a living system. The union of all of them is so much more than its parts that as a result you arise as an emergent being with all your self-awareness, conscious doings, intelligence, and wonderfulness."

A quick smile hushed over Carmen's face.

Jack paused and then continued. "I imagine God as being analogous to that. We are one cell of him and all of us, which may include all life forms in the universe, give rise to this incredible emergent being that we call God. A life force if you want to see it that way."

"So, the aliens are part of *your* God, too?"

"I don't know."

"Then, God won't help us, will he?" she asked sorrowfully.

Jack remained silent.

It had been two months since the alien encounter. *Deep Explorer* was repaired to the best extent possible. The starboard aft thruster was still inoperable, but all other systems were functioning well enough. Garibaldi and Pablo were even able to rewire and rebuild the heavily damaged cockpit. They were getting together in the meeting room to brainstorm about their next steps. Jack started with discussions of their logistics situation.

"Our fuel reserves are nearly gone. We only have enough fuel for course corrections and a little bit of additional ignition thrust. We only have enough food left for about six months. The most logical course of action is to return to Earth, get our ship fully repaired, and then join in the battle."

"We may gain some insights by sticking around a bit longer and to observe what the aliens are doing. We are quite safe here," Vladimir commented.

"We have been monitoring them for eight weeks. What do you expect to find that is new? If we do not return soon, this will be a suicide mission," Pablo said.

"These aliens are very different from us. It doesn't matter what we expect. They may surprise us completely. And this is pretty much a suicide mission anyway, even if we manage to return home safely. Or do you really expect Earth to put together a fleet in a few months that matches our adversaries? Surely they can adapt to our lasers…. The aliens have been around a long time, successfully!"

It was quiet for a while. Then, Jack spoke.

"Will see what headquarters says. If staying here and starving to death would save Earth, then I'd do it. We all would. But I believe they want and need us back on Earth. Every ship may count in the coming confrontation. And I don't think you give our species enough credit. I have faith in human ingenuity. There are nine billion people on Earth and I believe that the aliens won't find it so easy to stomp us out."

Later that night Jack talked to Mike Hang. "We have our ship as ready as we can. If we want to be back at Earth in time, we have to decide soon. We won't have enough additional ignition thrust left if we stay much longer. Our fuel sources are too depleted. Otherwise we compromise our maneuvering ability in any possible future firefight."

"I'll talk with the council and let them know that we need to decide."

"How ready are you on Earth?" Jack asked tentatively.

"Well, our best estimate is that we will have eighteen ships like *Deep Explorer* in space to face them by the time they are here."

"That won't be enough, Mike. Not by far."

"I know. But perhaps we can reduce their numbers enough to force their retreat," Mike stated but his tone betrayed his doubt. "There is one encouraging development though. We have engineered a crystallized quartz material that should be able to reflect their beam that destroyed the Void on Titan. Our first tests are very encouraging. We will also ship a prototype to Mars. Morty requested it. I think he is concerned that they are going to destroy Martian life first."

"That would be consistent if their goal is to eradicate all other life in the Solar System," Jack agreed.

"We are hoping for that. Either they will have to divide their forces or else slow down to attack Mars, which would give us a little more time. Both plans would aid us here on Earth."

"I don't know how much that will help," Jack countered skeptically.

"You might be right, Jack. But you showed us that we can destroy their ships. This motivated everyone. You gave us hope. The only other thing we can do is to make sure that they cannot defeat us from orbit. If they actually have to come down to the surface, I believe we have a fighting chance. Our fleet will intercept them before they get within firing range of Earth. Should they get through the fleet then we have a defense network of satellites ready in orbit around the planet. Each one contains twenty mini-nukes with auto-targeting and evasion software. We won't be defenseless."

"Maybe we should stick around here and try to inflict some damage on their home world?" Jack suddenly threw in.

"I don't see what you can do. The aliens are well defended there. You have low fire power, the Void is well hidden below the surface, and your crew is tired and exhausted. You have done enough. What else can you possibly do? You'd be destroyed before you even get close to them."

"I see your point," Jack sighed. He had already drawn the same conclusion but he had wanted to hear confirmation from Mike that his thinking was sound.

## 6.2 Change of Plan

The next morning Jack instructed Garibaldi to lay in a course that would bring them back closer to Saturn for a final pass to monitor their enemy and then use Saturn's gravity to slingshot them back to Earth.

"What do we do if they try to intercept us?" Kitahari asked in their morning briefing.

"We adjust our trajectory to avoid a firefight and use an alternative trajectory," Jack responded. "We shouldn't have any trouble evading them if they come after us, but we would sacrifice some of our fuel reserves and delay our return to Earth. But I think we need to try it. Our previous experiences tell us that they don't consider us much of a threat if we don't come too close. Of course, that may have changed. But it's worth a try."

"I will get the imager ready to take the best close-up of Tethys II that we can possibly get," said Vladimir.

"Very good, then let's get to it. Time to head for Home!" Jack said confidently. There was something uplifting about actually giving the word to return to Earth. It was obvious in the faces of the crew as well. They had been in space for a long time and the thought of going home, even under threat, was very gratifying. They laid in the course, fired the port board thrusters, and turned toward Saturn and its satellites.

"What's wrong?" Carmen asked Jack after the briefing.

"What do you mean?" he responded coyly.

"Something seems to be bothering you."

Her concern was apparent on her face, but Jack wasn't sure whether it was concern for him or something else. Of course he was troubled. How could he not be, but talking about it right now wasn't appealing to him. He just wanted to avoid a discussion, but maybe a brief conversation would placate her concerns. "I doubt whether we made the right decision."

"You think we should return to Earth on the shortest way and skip the imaging?"

"No." He said simply. He was trying to avoid the details, avoid showing his fears and indecision.

"What then?"

"Maybe we should take them on."

Carmen's jaw dropped incredulously. "This would be suicide!" she exclaimed, shocked at the mere suggestion. "And it's such an outrageous and typical male suggestion."

"I know," he replied, trying to calm her. "It doesn't make sense. But then—what we are doing doesn't feel right either." Jack was at a loss for words to describe his thoughts. Carmen looked expectantly for a response and for some justification of his idea. He took a deep breath and sighed before speaking again. He didn't want to upset her, but he didn't have time to adequately convey his feelings verbally. "Never mind, I'll try to sort it out later," and excused himself.

They were on their way home, but the initial excitement they felt had evaporated quickly. Everyone knew that there might be nothing to go back to. Worse, if Earth's Void were to be destroyed, they would probably all be dead no matter where they were located—like the Titan slugs on their ship. Vladimir and Pablo worked on the imager and were able to obtain very high resolution images. There was no alien reaction to their change in trajectory or even a hint they had even noticed it. Evening approached and they had another thirty hours until their planned sling-shot maneuver would direct them toward Earth. Pablo stayed awake this night and observed Tethys II through the imaging equipment, while the rest of the crew tried to rest.

Carmen lay next to Jack and he wrapped his arms around her until she fell asleep. He tried to sleep as well but his mind was restless. He tried to calm himself, but the feeling of unrest had grown stronger and stronger during the day, since he felt it the first time at the briefing. He sifted through their options again and again in his head, but there was no other logical solution. Finally, he fell into a disturbed sleep. In the surreal fog of semi-consciousness he dreamed again of the yellowish light that drew him to the Saturnian system. And lingering in the lonely oblivion of his thoughts there was also the dark entity from his dream. He tried not to face that sentient, searching consciousness, but he mustered his courage and searched quietly in the ether of his dreams.

He woke up breathless and shaken, but mentally calmed himself and quickly fell asleep again. The feeling of unease grew even stronger during the night. He woke up again. He felt his body internally vibrating and his limbs shaking. This time he had no memory of his dreams, but the residual fear they instilled was still festering in his stomach. All the answers seemed to be buried deep inside that alien planet. There was no way he could head back to Earth—that much was clear. Something, fate…instinct…desperation, drew him to the alien planetoid. But could he, should he, sacrifice his whole crew for a dream, or a bad feeling? Would they even follow him?

No, he had to find some other way. He would sacrifice himself. The others surely would think that he had gone crazy and that he only came up with this suicidal plan in desperation. Carmen would be totally upset. But that's what he had to do. Before his training at the Complex, he would have just dismissed dreams and feelings, but now the way he handled his subconscious was different. No, *he* was different. Everything in his guts told him that this was what he needed to do. He would confront Carmen first thing in the morning, and then the rest at the scheduled briefing shortly afterwards. His mind was made up, so Jack allowed himself to sink into a restful sleep.

When Carmen woke up, Jack was placidly sitting on his chair observing her and sipping his coffee. Carmen peered up from her pillow, bleary eyed and still half asleep. Her hair was tousled into a chaotic arrangement of frazzled tufts and the left side of her face now sported a trio of sleep lines that the bed linens has imprinted into her soft skin. Jack smiled involuntarily. How beautiful she was!

"Wow, you are up early!" she mumbled as she sat up. "What are you doing?"

"I'm just looking at you and admiring your natural beauty."

She smiled and then got serious.

"What's up? You look troubled."

"I am," Jack said simply. He hated to blast Carmen with upsetting news so early in the morning and had mulled over just how he was going to approach her about his decision. "I need to go down to Tethys II and try to destroy their Void. I will take the shuttle, while you return with *Deep Explorer* to Earth." Damn! He hadn't really meant to use the blunt approach. Jack watched as his words took a few seconds for her to assimilate in her groggy state. She sobered up fast as the implications of his statement took form in her mind.

"What? How do you think you will destroy the Void?"

She was frightened and growing frantic as fear took hold. Jack put up his hands to calm her. "I have no idea. The only thing I know is that I have to go."

"What about us?"

"I will always be with you." Jack's words sounded overly dramatic, even to him, and he was sure they were a small consolation to Carmen's fear of loss.

"Yes, as a dead person," she went red in her face. She just stared at him.

"Look," he said. "I need to do this. I can't find any peace otherwise."

"What about me? Us?"

"You know that the survival of Earth has priority. If there is only a sliver of chance I have to go for it!"

"Then I will come with you."

"No, I want you to be safe."

"What crap is that? I can take care of myself and have proven it several times in battle! Why don't you want to have me with you?"

Jack realized that this was a stupid comment and avenue to pursue. He hesitated. He rubbed his forehead. "Yes, you have proven yourself in battle. And that's why I need you to command the ship back to Earth. There is no one more qualified!"

"Oh sure, Garibaldi would be at least as well qualified. At least that way we could die together when it comes to it."

Jack realized that he really did not have any good arguments to sooth her roiling emotions. Of course, he had the right to make a command decision without the need to provide any justification. But he didn't want to pull rank with Carmen. "I'm sorry. I simply feel deep inside that this is the right decision, however difficult it is for both of us."

"You are just plain nuts!" She got dressed quickly and went out of the room, avoiding eye contact.

Jack just let her go, he needed to give her time to calm down and absorb the plan that threatened to steal him away; her strength and security. He sat back in his chair. He silently cursed their fate. He had found such a great woman and in another time and place he knew that he would have finally had such incredible happiness with her. He would have the type of love and serenity that poets write about. As sour as he felt about his doomed situation, he inwardly berated himself for his own selfishness. Compared to the survival of the human race, his lovesick tantrums seemed so petty. No, it wasn't fair, but he could not afford to pass up even a slim chance to save Earth.

He forced himself to calm down and let go of the things he could not change. He let his mind drift to happier thoughts. He truly enjoyed the last two months with Carmen. It was the best time of his life even given the dire situation otherwise. And he knew Carmen felt the same way. She told him so several times and she was clearly happier than on the way from Mars to Titan. Even her outer appearance and skin sparkled more than usual. And she also had experienced a bigger appetite lately. But now he had to let this wonderful person go. His thoughts became dark again. He felt cheated by life. In some way, their future was already taken away before it really started. If the aliens couldn't be stopped, it would be over anyway. And Jack doubted that they could be stopped by Earth. If there was any chance, or trace of a chance, he had to go down to their home world and try to stop them. Didn't Vladimir always say to expect the unexpected? Maybe there was some negotiating that could be done. Or maybe he would discover a vulnerable, critical part that he could destroy. The chance was vanishingly small, sure, but it would be zero if he just returned to Earth with the others. He needed a plan. But this would involve more crew members than just him alone.

At the morning meeting Jack explained his thoughts to everyone. He peeked over at Carmen. She was still red in the face and angry. His crew remained silent for a while, and then Garibaldi asked, "How do you intend to get to the subsurface access tunnels?"

"I thought about that," Jack said, "and have a rudimentary plan. We seek shelter with the shuttle in Saturn's rings. They should not be able to follow us given their bulky ships and slow reaction times. From the rings and Saturn's atmosphere we can extract Helium 3, purify it, and build a fusion bomb. We have all the equipment available on *Deep Explorer* and just have to move it into the shuttle. This will be a tedious task and will take weeks or months, but once we have succeeded we have a powerful weapon. Then we make a run for one of the access tunnels and get in."

"I'm delighted, Jack, you are being your old self again!" Vladimir proclaimed excitedly. "What the hell! If we are going down, then it will be at least in a blaze of glory," he laughed.

"We still have to work out some of the details," Jack cautioned, but he was pleased that he had such strong support from Vladimir. In fact he counted on it, because he knew that he wouldn't be able to pull it off by himself. Vladimir's ability to laugh in the face of death was reassuring but also somewhat disconcerting. Russians seemed to have such a fatalistic mentality, but Jack wondered if his old comrade really was crazy.

Garibaldi wrinkled his forehead and then said, "I think this is a feasible plan. I will go with you. You might need an engineer. Not to mention someone to watch your back."

Jack looked to Carmen for her thoughts hoping that the suggested plan and support from Garibaldi and Vladimir would calm her unease of this decision.

"You know my opinion," Carmen said calmly but her posture and tone spoke of anger and hostility.

"I'm ready to go down with you," Kitahari said. "The chances are as bleak on *Deep Explorer* as they are on that invasion mission, by my reckoning. We'll probably all die quick meaningless deaths, but at least out here we have a chance to do something. I guess the real question is: Do we have a realistic chance to make it against all odds and hit the jackpot? If the answer is yes, let's go!"

Jesus! We're all becoming fatalistic, Jack thought. Or maybe just desperate and reckless.

Pablo remained quiet and just nodded.

He is still young, Jack analyzed. He still has hope about returning home and isn't ready to face such a risk.

"Believe me, this is not easy for me. If it wouldn't feel right deep inside, I would never have considered it." Jack paused as he gathered his thoughts.

"Vladimir, Garibaldi, Kitahari, and I will take the shuttle and leave *Deep Explorer* at our closest encounter to Saturn, before you," and he looked at Carmen and Pablo, "get the gravity assist. We will try to infiltrate Tethys II. Carmen, you will be in command of *Deep Explorer* on the way back to the inner Solar System."

"We have a lot to do," Jack continued. "Vladimir, I want you to re-visit our improvised weapons ideas, especially the man-to-man combat ones. Also, we have to make sure that we have all the equipment for the Helium 3 plan. Garibaldi, I need your input on that, too, and you have to double-check all systems on the shuttle and make it ready for battle. Not for a head-on confrontation, but we need to avoid getting blasted. Kitahari, you take care of the logistics. Group dismissed. I'll inform Earth about our plan."

Everyone left, only Pablo remained and stood to address Jack.

"Kitahari should go with *Deep Explorer* and I with you," he declared blandly. His tone was resigned, Jack noted. Pablo's intent was probably to be chivalrous, but Jack questioned his commitment. Not that Jack questioned the young man's bravery or devotion to duty. Conveniently, Jack had legitimate reasons to exclude him from the mission.

"No, Pablo. There has to be an engineer on *Deep Explorer*, especially after all the damage to it. After Garibaldi you are the most qualified. I need you on the vessel." And also to look after Carmen, Jack added silently.

Pablo nodded his affirmative. "Yes, Sir," was all he said and then left the room quietly.

Then Jack went to the communication consol and initiated a telecon with HQ. Both Colonel Haggerty and Mike Hang appeared on the screen. Jack filled them in on his plan.

"You are crazier than I thought," Mike smiled. "What will you do if you actually get into their subsurface access tunnels?"

"We'll improvise. As Field Marshal Rommel once said, 'No plan survives initial contact with the enemy!' Since we have no idea what to expect we'll just have to rely on luck and ingenuity."

"I was afraid of that. What do you think, Colonel?"

"I do not like wasting equipment on a fool's errand, and throwing away human lives I like even less. But I think it may be worth a shot. We will still have the ship back in time with Carmen Mejilla as commander. She was in battle before and her experience will be valuable."

"Since you feel so strongly about it, you should do what you feel is right," Mike concluded.

"Who will command our fleet?" Jack inquired, relieved to have their support.

"General Bicker."

"Oh, oh." Jack was puzzled by the decision to grant command of the fleet to Bicker. Even more appalling to Jack was the thought that Bicker would have power over Carmen if she returned with *Deep Explorer* in time for the anticipated battle. But it was clearly not his place to demand a change in command at that level!

"He will do fine up there," Mike tried to alleviate Jack's concerns. "Besides, I expect the main battle will be on the surface."

"Is the Void sufficiently protected?" Jack queried with concern.

"Yes, we have several protective layers and a whole army around it commanded by our old friend, General Dahai."

"General? Hmm… but it makes me feel better. Dahai will make a good and wise leader."

"I need to attend to other duties, so I'll be signing off. Good luck, Jack, with your plan. You will need it."

"I know. Over and out."

That would likely be the last time that Jack would be in contact with Earth HQ. He sat back placidly. It had surprised him how easily they went along with his plan. But then, the situation was desperate and his presence really not required. Carmen was battle-proven and would do as well as he could in the captain's chair, or perhaps better. Why not take a chance? It should not make a difference where he would die. Either way, in the space battle over Earth or in his planned invasion, the survival chances were near zero. The only thing that bothered him was that Carmen and he never had a chance to build something together, to have a future, and possibly even a family.

Suddenly Jack noticed that Carmen was standing in the doorway, silent. Jack looked up at her with a questioning face, not sure how she would greet him. "Why? Why does it have to be like this?" Carmen was calmer now. Her anger had shifted to fear and sorrow. It felt like she mourned his death already.

"I don't know," Jack shook his head slowly, sharing her frustration with the path before them.

"I was truly happy for the last two months," she said as she moved over to him and embraced him tightly.

"Me, too," Jack mumbled into her hair. "Even though we are in this situation, it has been the happiest time of my life. I know it is against all odds. But as long as there is a trace of a chance, we've got to try."

"But if I die, I want to die with you and not alone in space."

Jack sensed that she had already accepted their fate and had come to terms with the likely outcome of this bold mission, but he needed to reassure her nonetheless. "There are larger issues involved here. You know that. *Earth, and all life on it.*"

"Yes, but why can't I go with you? The chances are about as bad in the battle for Earth as they are on your suicide mission."

"True, but you have choices. You don't have to engage the alien ships together with the fleet commanded by General Bicker. You can attack and then run as we did before! Or, depending on how things are going, you can land on Earth and hide, or go to Mars, or retreat to the Venus Orbital Station." Jack was grasping for any thread of logic that could reassure her.

"But if Earth's Void is destroyed, then all is over and it doesn't matter. You saw what happened to the slugs!"

"We don't know that for sure. And I don't truly think that Dahai is going to let that happen. He is commander of the armies around Jerusalem. I know him. He won't fail us. He doesn't know how to fail."

Jack was not sure whether he believed that himself, but it would make Carmen feel better, so he had to say it.

"Let's hope so."

Jack moved toward her and hugged her. Then he hugged her intensely. Tears filled her eyes and then his. Then Jack hugged her again, kissed her on her lips, took her by the hand, and went into the hallway.

Vladimir crossed their path in the hallway. "Look, what I found. Our new weapons!" he said excitedly, brandishing daggers, sabers, and a mace.

"What?" Jack exclaimed, more than a little dubious.

"Sure, I went through the personal assets of our dead crew members for possible things we can use and Zadak had a whole collection! He must have made these from junk or spare parts on board. I didn't know he had this kind of a hobby!"

"Will they be of any use?" Jack asked, slightly aghast. First they threw a high-tech rock at the aliens, then lasers, now back to big nearly pre-historic knives?

"Yes," Vladimir exclaimed enthusiastically, "they beautifully pierce their skin. I tried it out on the specimen we have in our lab."

"Do we have anything more sophisticated?" Jack said, still doubtful.

"Garibaldi put together a pretty powerful laser, but it's heavy, about fifty kilos, and can only be used by two people working together. Pablo found some iron rods that he is sharpening at the edges and we can use those as spears."

Carmen just shook her head, muttered "SPEARS!" in disgust, and walked away toward the cockpit. Jack looked at Vladimir and said "Jesus! We are in the second half of the 21st century and the best weapons we have are medieval? Or even stone-age?"

"No, no, they are not bad at all. These creatures probably don't have anything better in man to man combat. Look at their claws. They probably don't use anything else. And remember we can move faster. Even with suits on we'll still be more nimble and quicker than they could ever hope to be. And with all of our martial arts training we should make an impressive force. Hell! I bet I could take on any two of the buggers. With one hand tied behind me, too!"

"I'm glad you are upbeat," Jack smiled and clapped on Vladimir's shoulder. "That's the attitude I like. Barbaric, to be sure, but effective. We'll kick some alien butt." Jack didn't want to dim his enthusiasm by pointing out that if they managed to penetrate Tethys II, and this was a big 'if', they would probably have to face hundreds of aliens. He changed the topic. "What are the aliens doing anyway?"

"No action at all. They seem to be ignoring us, but I have no doubts that they are monitoring us very closely. We should be able to decouple the shuttle as planned."

"We need to finalize our trajectory after we separate from *Deep Explorer.*"

"I was about to do that."

"I will join you soon," Jack reassured him. "Let me first check on the others."

Jack looked for Garibaldi who was still assembling the laser gun with the help of Pablo. It was impressive but heavy. Garibaldi explained to Jack that there would only be enough energy in the power cells for three powerful focused laser blasts. Kitahari packed dried food, water, and their spacesuits into the shuttle. Apparently she had already met Vladimir since she had one of his daggers dangling from her hip. Then Jack joined Vladimir in the imaging room and discussed which trajectory to take and which of the subsurface tunnels to approach.

It was one hour until closest encounter and shuttle separation. Vladimir, Jack, and Garibaldi checked and double checked that they had everything in the shuttle they needed. Meanwhile Kitahari had brewed some liquids together to produce little mini-grenades and stored them safely inside the shuttle. Pablo stayed at the imager, but there was still no reaction from the aliens to their approach.

"We have twenty minutes until separation," Vladimir said.

"Oh, it's already that late? You go ahead. I still have to talk to Carmen." Jack ran to the cockpit.

"Are you ready for decoupling?" she greeted him.

"Yes, in less than twenty minutes," he replied somewhat out of breath from his haste to see her. "Are you ready to command the ship back to Earth?"

"Yes, I guess so."

Silence followed as they stared softly, expectantly at one another. Their words were hollow. There was so much to say and so little time to say it. Never enough time! And how could mere words describe their emotions or sooth the fears and pain of parting. There was no need really, their eyes and faces said it more clearly than the great writers of mankind could ever do.

"I'm sure we'll see each other again," Jack tried to reassure her.

"I truly hope so," she said with a depressed smile and looked away. She didn't want to face the loss. Then he held her. And for a brief moment he lost himself in that warm, tender embrace. He burned that feeling of love and serenity into his consciousness. A little slice of heaven to take with him into hell. It felt like a small eternity before he released her, kissed her, and left. Did he tell Carmen he loved her? Had she said the same? Jack couldn't remember. The details of the moment were a blur. But that feeling, that glorious feeling of her soft body next to his was not. He would never forget that.

The rest of his exploration team was already waiting in the shuttle for him when he scrambled in and buckled himself into his seat.

"All systems ready?" he asked.

"Aye," Garibaldi responded.

He heard Carmen's soft voice through the intercom: "One minute to separation," and wondered whether he would hear her voice ever again. Garibaldi had only been able to repair the short-range communication device in the shuttle and they would soon be out of communication. He resisted the temptation to call everything off just to have a little more time with Carmen. That was so pathetically selfish…, but also human.

"30 s…10 s… 9… 8…7… 6… 5… 4… 3… 2… 1… Release!"

The shuttle released and catapulted toward Tethys II. A few moments later *Deep Explorer*'s thrusters ignited, and changed course for the planned slingshot maneuver toward Earth.


## 6.3  Separate Ways

*Deep Explorer* sped towards the outer edge of Saturn in a Deep Insertion Gravity Assist maneuver, one that had never been attempted with a human before. It was dangerous. The strain placed on the ship and crew was acute. Carmen was confident they could make it, but the ship had taken quite a beating. The strain could rip it right apart. 'Love and duct tape', that's what Pablo said was keeping the ship together. But they didn't actually have any duct tape, so that left love. Carmen swallowed hard and fought back tears…Love.

*Deep Explorer*'s velocity began to increase. Carmen already felt the G-forces pulling her a little deeper into her seat. She took a deep breath and forced it out again. And then another while she still could. Pretty soon the simple act of breathing would be a Herculean effort. They would be pulling nine Gs at maximum. Nine Gs… for six minutes. That scared her a bit. The human body could withstand nine Gs for about thirty seconds without a G-suit before passing out. Any longer and the oxygen starved brain would start to die.

They were passing three Gs now.

Fortunately, Carmen and Pablo both had G-suits on. Carmen just hoped hers was functioning correctly. It didn't feel like it was working though, and a trickle of fear crept into her mind.

*Five Gs.* Carmen took another deep breath. It was a little difficult this time. Was the G-suit working? The ship began creaking a bit as it adjusted to the increased strain.

*Six Gs.* She blamed herself for not triple-checking her suit. Was Pablo okay? Her mind wandered—was this the onset of anoxia? She tried to lift her arm to toggle the intercom. But the force was too great and pinned her arm back to the chair. It slid off and pulled her arm straight down. She felt her blood rush to her hand and it swelled until it felt like a balloon ready to burst. Her fingertips tingled. She tried to lift her arm up but couldn't.

*Seven Gs.* Over the pounding in her ears Carmen could hear the ship groaning in complaint. It seemed so far away. Her hand was starting to hurt. Had their calculations been accurate? The smallest error in trajectory, velocity or thruster timing would destroy them.

*Eight Gs.* Carmen tried to take a breath to calm herself. It was very hard. She wasn't getting enough air. She fought off her panic. Just concentrate on breathing…that's the key. She sucked in air in short labored gasps. The ship was shaking like one of those cheap amusement park coaster rides. Those rides often broke and people would get hurt. She always loved riding them anyway. But not this ride.

*Nine Gs.* Moment of truth. She couldn't feel her hand any more but it felt like she was holding up a four hundred pound block. And the strain on her shoulder felt like someone was trying to slowly rip her arm off. Damn, but it hurt. Have thirty seconds passed? The ship was shaking so badly she couldn't even see her instrument display. The grating and popping of metal was too loud. There was a deafening metallic snap. And then three more in quick succession followed by the tinkling of shrapnel skittering across the floor. Was that a bulkhead collapsing? Darn!

The nightmare seemed to last forever. Until Carmen was sure that something had gone wrong and they were all going to die. But gradually the shaking and rattling of the ship lessened. It wasn't quite so hard to breathe now and she realized that she was sucking in air like she had just run a marathon.

After a few more moments she tried to pull her arm up. She couldn't. She had to use her other arm to pull her 'dead' arm back into the chair and onto her lap. It wouldn't move. Hopefully it would recover. She had to use her other arm to activate the intercom. "Pablo, report."

"I'm still here, Carmen. Wow! What a ride! I think we nailed it. We appear to be on a direct trajectory to Earth," Pablo said through the intercom. "We have gained enough velocity to catch up with the alien fleet somewhere between Mars and Earth."

Carmen sighed in deep relief as *Deep Explorer* sped toward the inner solar system, trailing behind the alien invasion fleet.

*****

Vladimir fired the shuttle only once to guide it on a wide trajectory towards Saturn's rings and Tethys II. It would take fifteen minutes for them to approach the outer rings, but Jack didn't want to attract any more attention than absolutely necessary. They would use the cover of Saturn as much as they could, but certainly they were exposed and could easily be noticed by the alien ships in orbit. But it was the best they could do.

Only a few minutes had past when several of the alien vessels broke their standard orbit and headed towards the shuttle.

"What should we do?" Vladimir asked.

"The game is afoot, Vladimir. Ignite the thrusters and move us into Saturn's E-ring. We have to get there before they do. We don't want to be at their mercy."

Vladimir did so and the alien ships pursued. It was touch and go as to whether they would make it or not. The shuttle didn't have much fuel for combat maneuvering and there was no guarantee that the ring debris would deter their antagonists. The shuttle shot toward the rings at a suicidal velocity. At the last possible moment Vladimir slammed on the retro-thrusters and decelerated. Even so Vladimir scrambled to safely maneuver the shuttle between the boulders of ice and rock at breakneck speed. And it wasn't just the big ones you had to worry about. At their velocity an ice chunk the size of a golf ball could blast right through the shuttle's hull.

Vladimir continued to slow and finally positioned the shuttle behind a large moonlet. It was hazardous since some of the boulders struck each other and were flinging out debris in all directions. The aliens closed the gap between them in no time. One of the alien ships tried to follow them to get a clear shot while the others angled around to drive the shuttle into the open, but a larger boulder struck one alien vessel in the side and caused visible damage. To everyone's relief it pulled out of the E-ring and retreated. The other ships retreated too and created a blockading formation around them.

"I guess there is an advantage if you can move and react fast," Vladimir said triumphantly.

"Yes, but now we're in check-mate." Garibaldi said. "We are stuck here."

"Not quite," Jack argued. "We're in check, to be sure. But the game is just getting started. Vladimir, use the ship's scanners to find the nearest and highest concentrations of Helium-3. I'll maneuver the shuttle deeper into the ring. Let's see how long they're willing to let us stay here in peace."

It turned out that the aliens were resigned to waiting them out. The aliens tracked and mirrored their movement whenever the shuttle changed directions or accelerated. But they held back from attacking. Most of their ships were very large, but even some of the smaller vessel did not dare to enter the rings.

If they knew what we were doing, I wonder if they would be so patient, Jack thought.

The days stretched to weeks as the crew collected every smidgen of Helium-3 they could find. There was never a significant accumulation of the rare isotope at any one location so they just had to keep moving, always wary of exposing themselves to the aliens. Too many times they were forced to bypass certain locations that would expose them to attack.

They were clearly wearing down, being cramped together in the small shuttle. Progress wasn't happening nearly as fast as Jack had hoped. Jack thought they would run out of oxygen before they could finish, but Garibaldi came up with a brilliant idea to increase their water and oxygen supply by harvesting water ice from the E-ring and splitting it into oxygen and hydrogen with the help of Saturn's electromagnetic radiation. They would be able to hang out here for several months if needed.

\*\*\*\*\*

Carmen held her belly. She'd suspected it the first time on that fateful day when she and Jack separated, but now it was clear: she was pregnant. She calculated the baby's due date. It was a little bit more than two months after their likely arrival to join the defense of Earth. Carmen sighed. She would have wished to see the baby even if the baby was never meant to be born. She became depressed and spent long hours just staring into empty space. Pablo could only guess what she was thinking about.

She would tell no one about the baby, at least not yet. It would just raise issues about whether she would be fit to command. And she definitely would want to make the decisions on her own when they arrived at Mars or Earth. If she could just let Jack know about this, but there was no way for him to receive a signal, if he was even still alive. She sighed at that thought. She felt lonely, lonelier than she had ever felt in space. But she had to be strong. Sure, neither she nor her unborn child had a good chance to survive the encounter with the aliens. All seemed so dark and senseless. But then there was her maternal instinct. Even with the chances being one in a million or maybe even less, she had to be strong and do what she could. Carmen tried to occupy herself with whatever she could think of. She read every book in the ship's data banks. Then she turned to working out. Then she turned to writing. But the depressive thoughts came back, over and over again: Where are you, Jack? Her tears made for bitter company in the cold, quiet expanse of space. She had to hang on, like on the trip from Venus to Earth after she got so badly injured. It was not clear to her anymore which was worse—the extreme physical pain she endured on her trip from Venus or the emptiness and wounded soul she experienced now on her way back to Earth.

Pablo was pretty quiet on their trip home and spent most of his time checking primary and secondary systems of their vessel while conducting improvements here and there. At least he had his duties to occupy his time. Though, he, as well, seemed to go through depressive stages during their journey.

Five long months for the trip from Saturn to Earth. Or, four and three quarter to Mars. Ironically, Mars would be nearly the same distance given the planetary positions at their time of arrival. Although they used Saturn's gravity assist, there was simply not enough fuel left to accelerate *Deep Explorer* to the speed they had achieved on their voyage to Titan. But they would be fast enough to catch up with the alien fleet in time.

They kept in regular communication with Earth HQ and were updated continuously on recent events. Carmen read through the latest Earth intelligence report: the alien invasion fleet had arrived in the inner Solar System. Four ships had separated from the main group and altered course towards Mars, while the rest continued their trajectory toward Earth. At least the alien force was somewhat fragmented—that was good. Mars had received the quartz crystal protection for its Void and would be the first to test this part of Earth's defenses. Morty was right, Carmen thought, Mars will be first. They were getting close to the alien invasion fleet and had to decide whether to join Bicker's fleet or assist Mars.

"What should we do?" Pablo asked passively.

Carmen thought through the situation for a while. Not that she hadn't done that over and over again during the last months, but this time she needed to make a final decision. "We will go to Mars. We should arrive a little bit earlier there. Perhaps, we can get their attention and draw more ships away from the main fleet. Maybe even take one of those bastards down with us."

"That's against our orders. You shouldn't let your personal feeling affect your decision. General Bicker will be steaming!"

"I don't care," Carmen smiled sardonically. "That's what we are gonna do!"

Typical female decision, based on emotions, Pablo thought. But she was in command and she might be right that it was more prudent to assist Morty on

Mars. Pablo noticed that Carmen had gained weight, but it didn't occur to him that she might be pregnant. He thought that she overate because of depression. Although one had to be pretty desperate to overeat on the dry food that they had with them, but then Carmen appeared to be depressed quite often. I guess both Carmen and Jack are very similar in making independent decisions, Pablo said to himself. The ideal couple, both of them stubborn and fatalistic!

Pablo noticed himself becoming more and more cynical. But how could you avoid it, given the situation? Perhaps, it was just another defense mechanism by his psyche, trying to remain sane. The losses, the long space travel, and the pressure were just too much. This was more challenging than the witch cage at the Complex, no doubt!

*****

Garibaldi sat down next to Jack in the shuttle's rebuilt cargo bay. He looked like some wretched homeless man Jack had met once in Seattle, a man with a scraggly beard and wild hair. "It took a very long time, but we are finally ready," Garibaldi said as he loaded the fusion bomb into the launcher. It was always amazing just how small such a powerful device could be. A century ago, a 10 megaton bomb weighed less than twenty kilos, and this was smaller yet.

They were ready to finally leave the security of Saturn's rings and die in their escape attempt if necessary. They all agreed that anything would be better than being constrained in this tin can a minute longer.

Vladimir took the pilot's seat and prepped the shuttle for flight. Jack sat as co-pilot. Everyone else was suited and buckled up in the rear of the shuttle. Jack couldn't believe that it had now been more than four and a half months that they had been trapped inside Saturn's E-ring. They hadn't had any contact with Earth or *Deep Explorer* for that entire time. According to their calculations, the alien fleet should be about to reach Earth. There wasn't any more time to lose. It was now or never.

Jack gave the go-signal and they accelerated out of the E-ring toward Tethys II. It took a while for the alien ships to react as they were apparently surprised by the sudden change after months of inaction. Or, were they just inherently that slow to react? Either way, it did not take very long until the shuttle was surrounded by ships that advanced closer and closer towards firing range.

"Now!" Jack shouted in excitement and Garibaldi launched the helium bomb projectile. It ignited at a safe distance with an incredible power, blinding friend and foe with such a light intensity as if the Sun would have gone supernova. Vladimir steered the shuttle into the blast zone and zipped in between two large alien ships. The blast did not destroy any alien ships, but it had the hoped-for effect of completely disorienting the entire group of alien vessels. By the time the aliens realized what was going on, Jack and his crew were far past them and had a good head start on the way to Tethys II. As soon as the aliens realized the situation, their ships accelerated after them and other vessels changed direction to intercept them.

YES! They were going to make it. But they still didn't have much time. Tethys II zoomed into view as they sped toward its surface. The shuttle turned at the last moment and skimmed the surface as it quickly decelerated. The force of the stop crushed Jack's body against the seat straps and practically gave him whiplash. They still weren't slowing down fast enough, though, and it looked like they would overshoot their drop point. At the last moment Vladimir jolted the shuttle onto the surface and the extra friction helped slow the shuttle even more. They finally came to a rough, grinding halt close to their planned stopping point right in front of one of the subsurface access tunnels.

"Bull's eye!" Vladimir whooped. "I bet you thought I'd miss it."

There was no time for a response.

The crew grabbed their gear and rushed out of the shuttle as quickly as they could, carrying the laser gun and their backpacks of oxygen, food, water, and weapons. The entrance of the tunnel was closed. Garibaldi and Jack used the laser gun and melted a hole through the goo covering the surface. Jack looked over his shoulder, knowing that the alien ships would be on them in seconds. Kitahari crammed the hole with explosives and they scrambled away quickly. A second later the explosives went off and blew a hole into the entrance large enough for them to enter the tunnel.

Without delay they sprinted to the hole and climbed blindly into the dark unknown. Outside, several blue streaks hit the surface and the ground was shaking. Jack braved a glance out of the tunnel and saw parts of the shuttle zipping through the atmosphere: he just hoped that the aliens didn't realize that they had escaped into the tunnels.

"I guess retreat is not an option," Kitahari commented as one of the metal pieces of the shuttle flew into the tunnel and impacted close to her.

"Neither is failure," Jack said defiantly. "Let's move!" They turned on their flashlights and moved further into the tunnel.

*****

The four large alien ships that split off from the main group approached Mars. Three of them positioned themselves in a triangular pattern above the Martian Void at an altitude of about 10 km. The fourth vessel was moving toward the Martian Base Station: that single ship fired blue streaks toward the station and instantly turned the once-proud outpost of human ingenuity into a pile of scrap metal. Meanwhile, the other three ships combined their firepower and fired a white streak at the Martian Void. That continued for a few seconds with the Void emitting a greenish all-pervading color. Then, the whitish quartz crystal shield that was installed below ground but above the Void moved over the Void and sealed it off. The streaks were reflected and hurled back to the ships, which blew up in a spectacular explosion in the Martian night sky.

"Yes!" Morty and the Martian Base Station crew cheered as they observed the spectacle from a nearby lava tube cave.

Unfortunately, the surprise was not working as completely as had been hoped. The one alien vessel still hovering over the remnants of the Base Station noticed *Deep Explorer* approaching, then turned and accelerated toward it. Abruptly, *Deep Explorer* and the alien ship were racing toward each other at high speed.

"All available power to the laser," shouted Carmen through the intercom and Pablo initiated the laser.

"The enemy vessel compensated, dodged the laser beam, and then fired a blue streak towards *Deep Explorer*. The streak struck the main propulsion system and the fuel reserves exploded out the starboard side toward space and the vessel careened into a spiral.

"Damn!" growled Pablo through the intercom.

"Get the laser back on their ship!" yelled Carmen.

Pablo struggled to keep his footing in the severe shaking, but was finally able to refocus the laser back onto the alien spacecraft. Another blue streak emerged from the alien vessel arrived and struck the habitat ring, blasting off large chunks into space. But Pablo managed to keep the laser on target until finally a hole appeared in the front of the alien craft and they saw interior contents blowing outward.

"Continue the laser!" Carmen screamed through the intercom.

Carmen held her course and continued to steer toward the alien ship. Only in the very last tenth of a second did she veer to the lower starboard side to avoid a collision.

But she had waited a split second too long. The thrusters just didn't have enough power anymore to pull the ship out of the way. *Deep Explorer* careened into the alien attack cruiser with bone jarring, concussive force. It was just a glancing collision, but *Deep Explorer* spiraled towards the surface of Mars completely out of control.

"We are going down! We are going down!" Carmen shouted through the intercom and through the communication relay to Earth headquarters. She noted with grim and bitter satisfaction that the alien cruiser was going down too.

"Put your spacesuit on!" Pablo shouted back. "The ship's going to break apart!"

Carmen tried to level out and gain control, but nearly all the fuel had exploded or was lost into space. She only had minimal thruster control and *Deep Explorer* was approaching the surface much too fast. Carmen was heading for the northern lowlands with what little steering ability she still had. She could hear bits of the ship ripping off as she barreled through the thin atmosphere. Just before impact the strain on the ship tore it into two large pieces.

*****

General Bicker had nineteen Deep Explorer type ships assembled between the Moon and Earth. When the alien fleet was only two hours away, he gave the order for acceleration and interception. Bicker had his ships organized in three waves, while the alien ships arrived in a formation with one ship in the front followed by four, then nine, thirteen, and sixteen vessels, with another four ships at the tail end.

Bicker's ship was part of the first wave so he could optimally organize the attack. The two other waves were positioned at the trailing right and left flanks. As the first alien ship came into firing range, Bicker gave the command to fire the lasers. The alien ship rotated its hull as a defense mechanism and changed directions constantly and randomly. The vessels of the first wave found it hard to get a long-enough lock on the alien ship to melt a hole through its hull. Just as the group of four alien ships approached, they were finally successful. The four alien ships opened fire as soon as they got close enough and quickly destroyed two of the Earth ships. As Bicker's wave was struggling with the four alien ships, the second and the third wave of Earth's fleet approached from the flanks and joined the battle.

The aliens had learned well from their first encounter, and were now constantly spinning their ships on their long axes, so that human fire could not be held on a single spot for long enough to breach the hull. Bicker's strategy was clearly not working. The second and third waves of Earth's fleet were not planned to be drawn into the firefight with the second wave of the aliens. Their objective had been to engage the groups of nine, thirteen, and sixteen vessels still to come. Instead, they were forced to assist Bicker prematurely.

The Earth fleet was partially successful—it managed to destroy one more ship of the alien group of four and caused another to retreat with severe damage, before the nine ship group of the alien fleet came into firing range.

The Earth ships were faster, but overwhelmed by the fire power of the alien ships. They hurdled one blue streak after another at the Earth ships, many of which where already punctured by multiple hits. Blue streaks dominated the battle only to be interrupted occasionally by laser fire. When a blue streak hurtled through the fuel tank of Bicker's ship, the command ship blew up in a chilling explosion. Then the group of thirteen alien ships arrived and the remaining Earth ships were being destroyed or disabled so quickly that the conflict looked more like a shooting exercise for the alien ships than a battle. One Earth ship after another blew apart.

Finally, the highest ranking remaining officer, Admiral Nikata, commander of the third wave, gave the command to all remaining ships to retreat and re-assemble behind the Moon. Some of the Earth ships were already too damaged to retreat and were picked off one by one, and destroyed by the alien ships.

The result of the battle was devastating for Earth. The aliens had learned from their previous encounter with *Deep Explorer* and gave the Earth ships few chances to target their lasers for more than a fraction of a second on any one target, not enough time to burn through their hulls. The entire battle was over in four minutes. Only two alien ships were destroyed, but from the nineteen Earth ships fourteen were destroyed and the remaining five scattered in retreat more or less damaged.

## 6.4 Invasion

Shortly afterwards the alien fleet reached the Earth-Moon system. The established defense system on the Moon launched in coordination with Earth's integrated satellite defense network nuclear warheads at the alien ships, but those did not have any effect at all. The Moon Base was quickly destroyed and Earth's defense network was disassembled piece by piece. Then four of the alien ships separated from the main fleet and changed course toward Venus. The aliens remained true to their previous strategy, three vessels to destroy the Void on Venus plus one command ship, Dahai thought, as he observed the strategy of the aliens from his command bunker.

Then three of the alien ships advanced into Earth's atmosphere and formed a triangular formation above the Near East at an altitude of about 5 km. But Dahai was prepared. Huge laser cannons ringed the Void in Jerusalem and were linked to radar tracking systems. Two of the three ships were destroyed within a minute, and the third retreated. The aliens responded by sending ten battle cruisers into the atmosphere and firing a relentless wave of blue streaks at Earth's defenses. The laser cannons fired continuously and fighter planes were scrambling to intercept the alien attack ships.

It was a vicious confrontation. The aircraft with their conventional weapons had little effect as they were not able to pierce the hulls of the alien ships. They were just a distraction for the alien ships, like swarms of gnats buzzing around a rhinoceros. The alien cruisers ignored them and concentrated their attacks on the land bound defense systems. The firefight was brief. In seven minutes the alien ships succeeded in disabling the laser cannons but their success wasn't without cost. Six of their ships crashed to the surface in thunderous explosions of fire and twisted wreckage.

Another three ships assembled in the atmosphere to form the usual triangular configuration and initiated white streaks aimed at Earth's Void. After a few seconds of firing the Void lit up in a dark-yellowish pervasive light beaming in all directions. Dahai activated the quartz shield and as on Mars the reflected beam destroyed the three attacking ships. Ten more ships dipped into the atmosphere firing their weapons and plowing up the ground around the Void.

They are trying to take out the protective shield of the Void, Dahai thought. The ground shook around him from the intense bombardment. Even in the deep bunker the impacts were felt vividly. House-sized boulders were flying through the air and jumbled pieces of metal, flesh, and rock were everywhere.

Dahai remained calm. The shield of the Void was buried deep, as were many of the bunkers they had constructed in the last months. The tremors and shocks from the impacts were felt on the ground for miles around the Void from the merciless intensity and persistence of the alien barrage. Dahai sat patiently and confidently in the bunker deep below Jerusalem's Dome of the Rock, waiting for the aliens' next move. The air bombardment would fail, he confidently surmised. Soon the aliens would conclude that they will have to come down to accomplish their goal.

As if on cue, five alien ships quickly descended and landed on a nearby hill. Hoards of insectoid soldiers poured out of their bellies.

So the fate of man will be decided, Dahai mused. He waited for a few moments to study the aliens' mobility and intent. Then he quickly barked out commands to his subordinates to counterattack, thinking that for all the high technology available, it came down to this in the end—foot soldiers and hand to hand combat.

The slow moving aliens, in an unusual environment, were no match for the well-trained soldiers commanded by Dahai. The alien forces were being slaughtered as Dahai's troops exercised their defensive superiority. Thousands were mowed down as they marched relentlessly and fearlessly into an inferno of waiting death and carnage.

The airplanes and rocket launchers, so useless during the air battle, were now finally effective against the vulnerable aliens on the ground. They rained death from above. It was a slaughter of untold scope. Dahai began to fear that the alien attack would actually succeed, though. His troops were beginning to run out of ammunition. Just as the zero line came, the seemingly endless supply of insect warriors trickled to a stop.

Dahai's forces pushed the advantage and were able to destroy three of the landed alien ships on the ground, but two of them were able to retreat into the atmosphere.

A remaining force of about twenty-five to thirty alien vessels remained in Earth orbit out of range of the remaining Earth defense systems. Dahai could only wonder about their next move. He scrambled to strengthen his position and re-supply his troops. The aliens had only been testing their defenses, Dahai knew. If the enemy attacked again in force, Dahai's forces could be overrun. He needed more men, he needed more supplies, and most of all he needed more time.

*****

Meanwhile, Jack and his team were advancing further into the tunnel system. The walls of the tunnels amazed them because they could not determine what they were made of—some mixed organic-metal composition. The tunnels branched off several times, but there was no indication which route to take.

For now, Jack decided that they should stay together as a team. He decided to follow the path that would lead them deeper into the tunnel system. They were alert and ready for an encounter, but amazingly nothing arose to bar their passage. Garibaldi indicated to them that the density of the atmosphere was constantly increasing until it leveled off at about two and a half bars. The composition was fifteen percent oxygen, eighty percent nitrogen, and a trace amount of the gases argon and krypton.

"I think that it is breathable. I will take off my helmet," Garibaldi said.

Jack nodded in agreement.

Garibaldi took it off slowly. It took a while to adjust. Heavy, biting, cold air at minus forty degrees Celsius surrounded him. He indicated that he was fine despite the cold.

The others observed him and then took off their helmets as well.

"This definitely needs some adjusting," Jack said. "But it is a boon. We should proceed like this to preserve our tanks, but any sign of danger and our helmets must go back on."

The others nodded, and they advanced further into the tunnel system and passed several constructions that looked like power generators. Vladimir asked whether they should destroy them. Jack just shook his head and advanced further.

"I'm wondering whether they are tracking or monitoring us," Kitahari commented.

"I doubt it," Jack responded. "If they knew we were here, I'm sure they would try to kill us at once. It seems that luck is with us. Let's hope it continues until we find a suitable target."

They were approaching another intersection. The hallways were empty. Jack looked at his altitude reader, and then went straight ahead. The tunnel was widening with more and more constructions on their sides. Jack waved his invasion group to the side.

"No combat or even contact unless absolutely necessary. Carmen told us that they are very good in communication. They may be telepathic. I think there is a chance that they don't know where we are. They may think we are dead. Or, maybe they simply do not care. But as soon as we engage them, I'm sure they'll know where we are. We have to try to get as close to the center as we can, to do as much damage as possible!"

The rest of the team agreed.

It was cold, dark, and somewhat spooky, and Jack remembered the experience he had with the dark entity in his dream. Or, had it been a dream? Regardless, it didn't matter anymore. They were on a path of no return. Jack led the team further into the tunnel system, which became continuously wider and wider. No aliens jumped into sight, though. It felt like a wild-goose chase to Jack, with no real clear idea what he was looking for. He thought it was a miracle that they had even gotten this far undetected. And time was not their ally. But there was no time to second guess his previous decisions. They were inside the alien moon and they had to try to inflict damage, as much as they possibly could. That was their task, everything else was meaningless.

*****

When Carmen woke up, she lay on a stretcher in a cave. Danny was just bowing over her.

"Oh, look, sleeping beauty finally woke up," he said trying to be cheerful.

"Where am I?" she asked hoarsely, her voice cracking.

"Well, as you can see from the surroundings you are not in heaven, but you are not in hell either," he smiled. "You're still on Mars. You were lucky to survive the crash and we found you in time before you ran out of oxygen. I'll call the doctor."

Carmen felt like she was in a trance or hypnosis with a pounding headache. Voices and sounds bombarded her senses as if she were hearing through a metallic tube. She became aware that the doctor and Morty had arrived.

"How do you feel?" the doctor asked.

"Somewhat dizzy and I have a real bad headache."

"That's to be expected, but I think you will make a full recovery."

Suddenly fear flashed through her head.

"What is the matter?" the doctor asked calmly.

"How is the baby doing?" Carmen asked fearfully.

"Your placenta ruptured and you lost some blood. We are not sure how the baby is doing. It still seems to be moving, though. We are really not prepared for this eventuality. You should have at least informed us about your pregnancy."

Carmen became pale in her face. Morty sat down next to her and held her hand, while the doctor left. "You know, you are very, very lucky that you survived. The heavily reconstructed and shielded cockpit probably saved your life."

"The only question is for how long. How is the situation on Earth?"

"Earth is in a battle for her life. Our fleet was successful in destroying some of the alien ships, but it doesn't look very good. There are too many of them," he sighed. Then he added, "But Dahai is still holding the ground at the Void."

"Did you hear anything about Jack and his group?"

"Nothing. The tracking signal from the shuttle indicates that they apparently made it to the surface of Tethys II and following this we picked up many explosions on our long-range scanners. They might have been killed when they attempted the landing." He paused for a while and squeezed her hand a bit more tightly. "The shuttle was destroyed for sure. We haven't been able to pick up the signal anymore. But it is conceivable that they were able to slip into the subsurface before the destruction of the shuttle."

"How is Pablo doing?" Carmen asked blandly.

"I'm sorry, Carmen. He didn't make it." Morty replied gravely.

"Don't you have any good news?" Carmen said bitterly, as resignation flooded her face. She knew she should feel sad for Pablo, but the truth was she didn't feel much of anything. She felt numb.

Morty took her silence as an opportunity to continue. "Well, you are alive and so are we. For that we should both be thankful. I'm glad that I took Jack's advice and built an outpost in the caves near the Void. That's where we are right now. The Base Station is completely destroyed, but we'll still try to salvage as much as we possibly can. We moved most life-essential equipment and also all movable sensors into this outpost. We should be able to live on our supplies here for months. And if we can get at least a small part of our greenhouse workable again we could last for many more." Morty paused. "All of my men are alive. Only one was slightly hurt, slipping off a boulder when the aliens fired the streaks."

Carmen wasn't in the mood for relatively unimportant details. "Who is the doctor? I don't remember him from our stop on the way to Titan?"

"Oh, he was transferred from Earth three months ago and came with the material that was shipped to protect the Void. The crystal shielding worked very well, as you saw from orbit."

The doctor returned. "I think we have to give our patient some rest, Morty. She needs to get some sleep to recover." He injected something into Carmen's arm: she did not resist as black walls narrowed her vision and she fell into a deep sleep.

*****

Dahai hunkered down in the bunker as one impact after the other shook the ground above him. He was on a telecon with Mike Hang in Washington D.C.

"Situation report!" Mike yelled through the telecom link so he could be heard above the noise.

"We are holding," Dahai shouted back. "We should still have enough forces to repel another landing. But we are taking heavy casualties from this constant bombardment."

The bombardment had been going on for twelve straight hours now. It was relentless. His troops were exhausted and ragged from the onslaught. Many brave men had already succumbed to the stress. Some would just start screaming hysterically for the shooting to stop. Some would huddle in their bunkers and cry or rock themselves catatonically. The truth was, he needed fresh soldiers, more supplies and more ammunition. But with the constant attack he couldn't safely get any transports to relieve or bolster his position. "How is it looking in D.C.?"

"The aliens seemed to have changed their strategy. They are bombing all of the larger cities. New York, Boston, and Baltimore are already completely destroyed, and our radar shows them moving towards D.C. A second group of alien ships went to Europe. They have already destroyed Istanbul, Athens, Belgrade, and Rome. The cities are defenseless and there's not a goddamned thing we can do to help them. We should have built more lasers. Very few cities have substantial laser cannon emplacements around them. Our dead must number in the millions."

"Were you able to take at least some of them down?"

"One in Europe, and that's it. East of Istanbul, too, I believe, but that's not confirmed. They may have just driven it off. Someone launched a nuclear warhead from Baltimore against our orders, with the only result that half of the East Coast is now exposed to radiation fallout in addition to the destruction caused by the aliens. The command structure is breaking down, I'm afraid. The government has evacuated to deep bunkers, some around D.C., some in other parts of the country."

"What about the ships still in space? Aren't there some that survived the battle?"

"Yes, but the President wanted them in D.C. to defend the capital. I convinced him that it would be better to have them hide out of the reach of the aliens until a good opportunity arrives to help you defend the Void. None of our lives means anything if the Void is destroyed. That's where we need to focus our energies."

"I appreciate all this." Dahai said gravely.

"Don't give up, Dahai! The world is counting on you now. Fight to the last man. The alien fleet is arriving. I have to go"

"Good luck, Mike!"

"You, too, Da…"

The connection went suddenly dead. Dahai reflected on the situation. There wasn't time to even think about the rest of the world right now. Mike Hang was right: the aliens' destruction of the big cities was either intended to divert humanity's resources away from the Void, or they were pursuing a secondary objective if their first goal couldn't be achieved—to bomb humankind back to the Stone Age so they would not pose a future threat.

Earth's salvation rested solely in his hands and those of the brave men guarding this bomb-raked, forsaken stretch of scorched earth.

The shield was still in place and the aliens were apparently unable to destroy it from orbit. That meant that they would have to come down here again to meet him, on the ground, face to face. Sooner or later they would come. They just had to hold out till then—it was their only chance. But with the constant bombardment eroding his troops and resources he was beginning to doubt whether he would be able to push them back.

## 6.5  The Hive

Jack and his intrepid crew of invaders approached a tunnel outlet that led into a large room several stories high. They cautiously peered into the room. It was filled with gargantuan assemblages of machinery that overlapped multiple levels of space. Platforms, catwalks, scaffolds and support struts crossed and linked the area in a complex network. No control panels or monitoring stations were visible, but the machinery emitted a deep, throbbing hum.

"What do we do?" Vladimir asked.

"Let's set up the laser gun," Jack responded.

Garibaldi and Kitahari, who had carried the laser for the last stretch, set it down.

"What's our target?" Garibaldi asked.

"See the center area? That looks like a reactor. That seems to be a pretty good target," Vladimir suggested.

"There is another one like this, but a lot bigger, a few stories down," Garibaldi said, pointing to it.

"Yes, but how do we get a shot at it? The line of fire is obstructed," Vladimir countered vehemently.

"I agree with Vladimir," Jack said. "We will have only one clear shot on this. If they discover us, it's been wasted."

Garibaldi took aim.

"Fire when ready," Jack said coolly.

The intense laser beam hit the reactor, which was located on the center structural beam that interconnected different levels of the construction. A loud explosion shook the tunnel system and the center beam tilted crazily out of alignment. Several secondary fires ignited and a shock wave blew the whole team back into the tunnel.

They scrambled back to their feet. Garibaldi grabbed the laser and they ran back up the tunnel from which they had come, knowing for sure that the laser fire and damage would draw some attention.

They ran back to the last intersection and took a left turn rather than retreating straight back up their original path.

"Let's try for one more shot like that," Jack said excitedly, ecstatic to finally be making their enemies pay.

The ground was still reverberating from the calamity they had caused. They ran as quickly as they could down the tunnel. It split again, and they took the right fork—they headed deeper into the hive-like edifice. The entire structure must be enormous, Jack thought, amazed at the seemingly impossible feat of constructing something so massive. After several minutes the tunnel expanded and the illumination increased several-fold. They continued on down this major tunnel for about fifteen minutes.

Just as they were wondering if the entire planetoid had been abandoned, they approached an intersection at which two aliens turned into their direction. The aliens paused for a moment and commenced swaying side to side, their antennae writhing excitedly.

They raised menacing claws in the crew's direction. Garibaldi was about to use the laser, but Jack stopped him. "No, don't waste it! We can move faster than them! We go around them and use our other melee weapons for now. We save the laser for a more valuable target. You and I will run around them, while Kitahari and Vladimir take on the aliens."

The creatures approached in a ponderous, slow-motion charge. Their claws were wickedly serrated and intimidating, each over a meter long. The creatures themselves were about triple the size of their human counterparts. However, they did move much more slowly than humans. Jack and Garibaldi easily evaded them, even while carrying the laser, leaving it to Vladimir and Kitahari to confront them in battle. Jack and Garibaldi forged ahead stumbling through a tremendously wide and high opening. Through it they saw so much equipment, lights, and aliens moving about, that Jack and Garibaldi needed a while to overcome their amazement and identify an optimal target. The room was so huge that they could not make out the other end. Aliens were scrambling to and fro, each apparently intent on some particular task or mission. It was literally a 'beehive' of activity. Or maybe more like a glimpse of the interior of an ant mound from an ant's perspective. There must have been hundreds of thousands of the creatures in this chamber alone.

Kitahari and Vladimir didn't have time to contemplate Jack's strategy as he darted away from them. They drew their sabers and their home-made spears and charged toward the aliens, screaming war-cries as they went. Kitahari leapt straight at the first alien, turned quickly, evaded the claws that swung right through where she had just been standing. She moved toward the creature's head, evaded its pincers, and jammed her saber and spear through its head repeatedly. The alien collapsed and a whitish fluid gushed from its head. Vladimir charged forward, dove left as the alien's claws tried to grab him, spun and delivered a viscous slice into the alien's thorax. He was rewarded with white goo gushing outward from the gaping wound and the creatures entrails began sliding out. But the creature's legs were still moving, as if the alien wasn't yet aware that it was mortally wounded. The back end of one alien leg hit Vladimir in the chest and knocked him back. He slipped on the wet floor and dropped prone. The alien lumbered forward and loomed over him with its wicked claws.

Kitahari saw this and leapt as high as she could toward the alien. Her blade lashed out and struck the alien in the back of its bulbous head, slicing neatly through it—the top of the creature's head flipped away like a misshapen pancake. The alien collapsed in a heap half-way to Vladimir.

"Are you alright?" she asked frantically.

"Yes, my spacesuit is ripped, but otherwise I seem to be okay." He stood up, somewhat shaken. "Wow! That felt good! Let's go find Jack."

Meanwhile, Jack and Garibaldi were readying the laser gun for a shot at a huge console from which thousands of smaller compartments branched off. The console was ringed by several aliens that were much larger than most of the others in view. Their abdomens especially were grossly larger and at least ten times the size of those of the other aliens they had encountered so far. The aliens spotted the crew, and a huge number of them began moving towards them, coming from every direction.

Jack adjusted the laser. His focus on the target was absolute. He knew that there was only one shot left in the laser. He had to make it count. In his mind he merged with the target, very much like the approach that he had learned from some meditation exercises at the Complex. Everything around him, the strange environment, Garibaldi lying next to him, Vladimir and Kitahari coming to his aid, the hundreds of aliens advancing towards him, was only a haze. There was no past, no future, and no fear. There was only him and his task. Time appeared to stand still: he was living in the moment. Slowly, very slowly, he moved the trigger. When the trigger finally reached the end of its travel and the gun fired, it was like waking up from a deep sleep—it happened so suddenly, and threw him back into reality.

The laser beam hit dead-center on the console which instantly went up in flames. The flames spread rapidly, toward the countless nearby compartments.

The damage was so much greater than Jack had hoped for! "That looks like a nice chain reaction!" said Vladimir, now breathing heavily next to him.

"I'm glad you caught up and cleared our retreat. Let's get the heck out of here," Jack said as he turned and started running. They abandoned the laser and fled as fast as they could. Several explosions rocked the ground behind them and reverberated down the tunnel.

"Which way are we going now?" shouted Garibaldi, gasping for air.

"Whatever way leads up towards the surface!" Jack shouted back.

They took a right at the next intersection, went straight, and another right. They could only imagine how many hundreds or thousands of the creatures were behind them, trying to get them. But those aliens were all slower than the humans. Even in this alien, cold, high-pressure environment the crew quickly outpaced them.

Until they ran straight into a group of six aliens, three soldier types with pincers, and three worker types without. There was no way back, and no way around. The tunnel was small and they had to face their enemy. Kitahari attacked first with her saber and dagger drawn, followed by Jack.

"You have to pierce their heads" she yelled as she charged. The aliens were slow, but they were working in concert. It was a real challenge to avoid being grabbed by any of them. Jack and Kitahari were the most agile, readily avoiding claws and legs, and slicing into their alien adversaries as they ducked, dodged, parried, and slashed.

Jack silently thanked his teachers for all the hours of martial arts training that he had endured back on Earth. It was certainly paying off now. It took just a few moments to gauge his opponents—their reach, reaction time, attack and defense capabilities. Jack eased into the thrill of the fight. He sliced though one soldier's claw and then followed up with a back-handed slice through its head. The creature flopped to the ground and another lumbered forward to take its place. Jack glanced at Kitahari: she had already killed two workers.

Garibaldi stepped forward to aid Jack but he slipped on the white ichor that oozed all over the floor. The engineer fell, landing hard on his back on the floor. Jack tried to distract the soldier attacking Garibaldi, but the advancing alien towered over him and its claws were descending too fast for Jack to react. Instinctively Garibaldi rolled out of the way and blindly thrust upward with his spear and jammed it right through the alien's head. The alien collapsed onto Garibaldi, pinning him beneath several hundred pounds of dead weight. A flailing leg slammed into Jack and knocked him backward several steps and into Vladimir.

Before any of Garibaldi's comrades could help him, the last soldier-type alien moved in and with a sheering pincer movement simply ripped off Garibaldi's head.

Vladimir leapt past Jack. "Noooo!" he screamed as he threw himself recklessly at the alien. He jumped on top of it, driving his saber from the top to the bottom of its head. The alien collapsed and fell to the ground, so did Vladimir, right in front of the last remaining alien worker that Kitahari was battling. The alien worker inadvertently extended its claws directly into Vladimir's path.

Vladimir moved to the side to evade it, but a sharp protrusion on the claw went right through his leg. Vladimir screamed in pain. The alien worker turned towards him to finish him off and was about to use the claw of its third leg to kill him when both Jack and Kitahari sliced into the alien organism simultaneously. The instantly-dead creature collapsed into a heap.

Jack pulled the claw from Vladimir's leg and blood darkened his spacesuit. The Ukrainian only grunted but his face blanched.

"Go on! You have to run!" Vladimir shouted. "I'll slow them down for you."

"You stubborn fool," Jack responded. "I'm not going to leave you here. Even with you on my back I'm still faster then those aliens," he said resolutely, and hoisted him onto his shoulder.

"What about Garibaldi?" Kitahari said.

"Leave him," Jack said sharply—but it was not an emotionless decision. "There's nothing we can do for him." Jack didn't want to think about what horrible atrocities the aliens might inflict on his corpse, but there wasn't anything he could do about it. He could tell that Kitahari and Vladimir were hesitant to leave Garibaldi's body behind.

They heard heavy footsteps—aliens—in the tunnel they had come from. "Let's go," Jack ordered.

"We have to hurry!" Vladimir shouted in agreement.

Jack grabbed Kitahari by her right hand and pulled her. The conflict was apparent on her face for a second, but then she made her decision and ran with Jack further up the tunnel. After a few hundred meters Jack stopped to catch his breath.

"I think the air is getting thinner," Kitahari said, now clearly more settled. "We should close our helmets again."

"Right… good idea. I need some warmth anyway," Jack responded. "My face feels frozen."

"You know, I think we just killed their brood nest, or at least one of their nurseries," Kitahari said.

"You think so?" Jack smiled hopefully.

"Yes. Those chambers look to me an awful lot like incubators. And with all the big-bellied queen-like aliens around the console, definitely."

"No wonder they're so pissed!" Jack said happily.

They helped Vladimir put on his helmet. He was obviously getting woozy from blood loss. His suit would not protect him much from the cold, since it was breached at several places. Then they continued their ascent through the tunnel system. Jack tried to keep some sense of direction, but it was hopeless. He really had no clue which direction they were going anymore. Maybe they could find someplace to hide.

When they approached an apparent tunnel outlet they noticed that it was directly connected to some other type of construction. Jack wasn't sure if he should enter this new area or not, but suddenly he heard the sound of alien footsteps approaching from behind.

He rushed into the new area, followed by Kitahari. Jack quickly realized that they were inside one of the alien ships—it looked like a duplicate of the one that he had boarded at Enceladus. They set Vladimir down gently, then Jack and Kitahari moved in a bit further. They saw one huge alien with an abdomen at least ten times as large as the aliens that they encountered in the tunnel. Its head was also at least triple the size of the alien soldier and worker types.

"That must be one of their queens," Kitahari whispered.

"I think we're in the control room, but I don't see any controls," Jack growled in frustration.

"They may fly these things telepathically or using some other type of communication," Kitahari speculated. "That would be consistent with their physiology."

"What do we do now?"

"Okay, this is crazy, but I would suggest approaching that alien, one of us getting a weapon right at her abdomen, the other right at the head, but not piercing it. Unless it attacks us, in which case we kill it immediately" Kitahari said.

"You mean kidnapping?" Jack grinned.

"Yes, we'll take the bitch hostage."

"How will the creature know what we want? It is an alien after all—we have no idea how it will react or interpret our actions!"

"If it is intelligent, it should be able to guess our intent even if we can't communicate their way."

"Well," Jack said, "you are absolutely right. It is a crazy idea, but I like it. We can always kill it if it doesn't cooperate. I take the head, you get the abdomen."

Kitahari and Jack rushed forward. The alien queen moved her front claw, but was much too slow. Jack parried it aside with the flat of his blade and then set the sharp saber tight against the side of the creature's head—Kitahari had her dagger and saber touching its huge abdomen.

The alien queen seemed to understand her precarious situation. Her body went completely still, but her antenna moved frantically to and fro.

Jack gesticulated with his hands and arms, in an attempt to communicate their intent to leave with the ship but there was no reaction.

"The creature doesn't get it," Jack said.

"Try drawing a star map," yelled Vladimir from behind while he hobbled into the room.

"Good idea," Jack agreed. Jack took out his knife and carved Tethys II and Saturn, Jupiter, and the inner planets into the soft surface of the ship's console. Then he pointed with his knife toward the alien, then to Tethys II, and then he carved a line from Tethys II to the third planet of the inner Solar System.

The dark, big eyes of the creature revealed nothing. The antenna still moved wildly. It was impossible to read this creature, and Jack wondered if it had the same problem understanding humans or if he was in fact completely transparent to the alien. Jack got exasperated.

Something was moving behind them: eight alien soldiers stormed into the room.

Jack got tense. He held the saber right at the head of the queen. Both Vladimir and Kitahari had their weapon directed toward the queen's abdomen. The alien queen moved her antennae rapidly, but held still otherwise. The alien soldiers stopped. Nothing happened for a few seconds. The moments of stillness and silence drug on.

Stalemate, Jack thought. He wasn't going to back down. No matter what happened they would at least kill the queen if the soldiers attacked. Then he would sell his life for as high a price as he could. He grimly set the blade of his saber a little harder against the queen as if to emphasize his seriousness. He relaxed mentally. Whatever would come next, he had already accomplished his mission as well as he possibly could have hoped for. He was at peace with himself and ready to die if necessary.

The stalemate continued for a few more seconds that seemed to last for an eternity. Then the alien soldiers responded and moved out of the room and left the scene.

So we CAN achieve some mutual understanding, Jack thought. Good. A healthy desire for self-preservation served them both well. As Jack contemplated his next step, he again carved the map on the console in front of the queen. Something was starting to move. Jack felt a tremor in the ship and then heard a screeching metallic sound.

"What is going on now?" Vladimir yelled.

"I think the tunnel is closing," Kitahari responded. Jack forced himself to be patient and stay relaxed. After all, these aliens were slow. Then the ship moved.

"I guess the queen got it," Kitahari added.

Jack was cautious. It seemed to be too good to be true. "It might just want to trick us."

"I don't think so. Those creatures seem to give their offspring the highest priority. And this queen must be carrying a lot," Kitahari said, as she pointed to the belly of the creature.

"Maybe we have something in common with them after all," Vladimir added.

## 6.6  The Dawn of a New Age

The front of the ship shimmered and became translucent from the inside and Jack, Kitahari, and Vladimir could clearly see Saturn and its rings as they moved away from Tethys II. It was a magnificent picture. They also noticed that all of the other alien ships were moving the other direction and landing on the surface of the planetoid.

"That's a good sign. I guess we did a lot of damage, and they are all helping out in repairing it," said Jack in a triumphant voice.

"It might be more," Kitahari added. "The mother instinct. The nest, the hive, is in immediate danger and all the adults are rushing back to save the young."

"Wahoo! It sure looks like it. We did it Jack! And we're still alive too," hooted Vladimir. Then they all fell silent. They weren't all still alive. Each of them privately mourned Garibaldi's sacrifice and absence.

*****

The situation on Earth was nothing short of catastrophic. Three quarters of all major cities were completely destroyed. The group of alien ships above Jerusalem was flying one sortie after another. The remaining Earth ships tried an attack, but could only take down one alien ship, while three of the remaining five attacking Earth vessels were destroyed. Dahai hunkered down in his bunker and transmitted to D.C. headquarters. Visual and audio transmissions were no longer possible so he sent a simple text message:

*This is our most desperate hour*
*I don't know how long we can survive here*
*The morale of our troops is near zero*
*And I doubt we can fight back another landing*
*If you still have any means to help us,*
*Please do, otherwise everything is lost.*

Dahai was not even sure whether anyone would be able to receive this transmission and even if they did, he didn't see how they could possibly help. He hadn't heard anything from D.C. in hours—not even from the well-protected command bunkers. There was not much that he could do, just protect the Void as long as possible.

A new report came in: the aliens were coming in again. Dahai waited for them to make their run. Then he waited some more. The attack didn't come.

"What happened?" he asked his staff, fearful that the aliens were trying a new, more devastating tactic.

"It looks like they turned back and their ships have moved back into orbit, Sir," one of his assistants declared.

"What for? A new landing? Come on—we must still have a functioning radar somewhere! I need to know what they're doing."

"Give me a moment, Sir."

There was a long pause. Dahai waited, his face an impenetrable mask of patience.

"It looks like they are all assembling between Earth and the Moon. No… they are now moving back towards the outer solar system."

"The attack is over? Are you sure? Can anyone confirm?" Dahai tried to contain his mounting excitement. Tense moments passed.

"Yes, Sir, General. We have confirmation from the Venus Orbital Station. They thought it was their last hour, being sitting ducks out there with no defenses. But even those alien ships turned around and are speeding back to the outer solar system."

"Why?" Dahai thought out loud. "That doesn't make any sense. We were done. They had us on our knees." He was skeptical. He wasn't ready to accept that the aliens would just quit when they were so close to victory. It just didn't make any sense.

After a few hours Dahai finally got back a transmission from headquarters. A tired and worn out Mike Hang appeared on the screen. "It seemed that we have prevailed."

"What happened?" Dahai asked, pressing for a reason for the blessed turn of events.

"We don't know for sure. We only know that all their ships turned back simultaneously. And from our few remaining high-power telescopes on the ground, we know that this occurred just after there was an explosion on Tethys II."

"How big of an explosion? Did it take them out?"

"No, very unlikely. There was a definite increase in light intensity, but otherwise the moon appears to be the same. If we still had our Deep Space Telescope functioning we could get more information, but that was one of the first instruments the aliens destroyed. So, we can only guess. We don't even know for sure that this is related. The closeness in time may be just coincidental."

"I guess it doesn't really matter at this point what the reason is. I'm just thankful to still be here. I guess we can celebrate. We have prevailed…. for whatever reason."

"I'm not so sure about celebrating. Most of the large cities are completely destroyed. Most infrastructure. The death toll surely tops hundreds of millions and may reach one or two billion."

"That bad? We were out of contact for hours." Dahai was solemn. It was hard to look on the bright side of things with so much death.

"Well… you are right. Whatever the death toll, Earth has survived and we will recover."

*****

Carmen was lying quietly on her bed when Morty and Danny stormed into her compartment.

"We won! We won! The alien fleet is retreating!" Morty was shouting excitedly.

Carmen breathed in relief. "What? What happened?"

"Everything is still sketchy, but they found somewhere on Earth an operating space vessel and put up a high power telescope into orbit. All of the alien vessels are retreating to Tethys II. The retreat occurred right after something blew up on the alien moon."

"Jack," Carmen smiled, hoping against all odds that he was still alive, that somehow he had managed to survive.

"We think so, too. Even better," Danny added, "All the alien ships are returning to Tethys II except one that is coming this direction. It must have launched just after the explosion given its speed and location. We think that it might be Jack on the way back home. In one of their ships! Somehow he pulled it off."

"Oh, my dearest God! I'm sure it is him!" Carmen said confidently. Then she lay back and grimaced.

"What is it? Is everything alright?" asked Danny, worried.

"Get the doctor. The baby is coming early. I have contractions, they are strong…"

"Oh…dear…the baby is coming!", and Danny rushed out of the compartment.

After a few minutes the doctor arrived. The delivery took two full hours, but eventually a healthy baby girl screamed out her first sounds. She was the first human born away from Earth, breathing her first gasp of Martian air. Carmen named her Marta. It was 0601 on 9 May 2063, Universal Time.

The dawn of a New Age.

# 7   The Voids

## 7.1   The Link to another World?

14 years later, Mars, Valles Marineris Canyon.

An orange hue was pervading the thin Martian atmosphere. The void had been fully freed from its surroundings of overlaying rocks and sediments. Several instruments were built around it and focused on any miniscule energy blurb the void might release. A spacecraft descended from the sky and reddish dust swirled around the nearby compound. It took a few seconds for the craft to set down. The door opened and a slender astronaut climbed out.

"Hurry up," it shouted from the inside through the intercom. "A dust storm is supposed to creep up on us in less than an hour."

"Don't rush me, Ken," the other astronaut replied. "The downloading of the data will take a while."

"I just don't want to get stuck here in this god-forsaken place!"

Tanya was puzzled how someone with that attitude could have joined the Mars Exploration Team. She shook her head while placing her foot on the hard reddish soil. Then she walked over to the compound, which was nothing more than a bunker built from the surrounding basalt. Tanya entered the compound, walked over to the instrument panel and initialized the data download. Then she left to check on the monitoring instruments. Yes, the wind was definitely gaining power, but there was no associated sound with it. Sound did not travel well in the thin Martian air. It was eerily quiet.

Tanya adjusted one of the scanners, which was slightly out of focus. It must have occurred from some earlier strong winds recorded in the valley. It was early afternoon, but the light diminished quickly. As the wind picked up, the air became loaded with red dust. After adjusting the monitoring devices she went back to the compound. The data download would still take a few minutes and she glanced over the records. She jolted as suddenly the door behind her opened.

Ken carried in some large box with supplies.

"What are you doing?" Tanya asked.

"The storm is picking up fast and furiously, and Daniel Perera decided that we should hunker down here until the worst is over rather than lift off during the storm. I told you to hurry up," he said with some tenseness.

Tanya was annoyed. Ken was probably the last person she would want to spend the night with at a remote location. She didn't like that bulky short guy very much with his not so subtle jabs at her race and upbringings. Her hometown of Houston had been completely obliterated during the alien attack on Earth and she was one of the few thousand survivors in a city of three million. She ended up in a foster home in Utah where her black skin color stuck out like an orange in a bed of green apples. But her foster parents were good to her and she began to love the barren rocks and salt flats in her newly adapted home and later set her goal to join the government-pushed effort to establish an expanded settlement on Mars.

This settlement consisted now of a total of 58 people, mostly scientists to explore the Red Planet, and engineers to expand the rebuilt Martian Base Station to a site for a more ambitious future colonization program. Daniel Perera, station commander for the last four years, was leading this effort. Nearly half of the scientists were occupied to collect and analyze data from the void, while the other half was divided in many different subgroups with various goals. She recalled meeting Jack Kenton, who was leading the group to grow plants on Mars and from whom she had heard so much.

"Don't you want to get your emergency box?" Ken ripped her out of her thoughts. "Sure", she replied and trotted over to the spacecraft. She probably wouldn't get much sleep anyway, but to haul some cushions and basic food supplies to the bunker wouldn't hurt. Who knows how long the storm would last.

The storm intensified and twilight set in. Ken made himself comfortable in a corner and started reading. Tanya felt uneasy and walked from the instrument panel to the door and back.

"Why don't you take a stroll outside if you can't simply sit down?" Ken blubbered.

"I'll do that."

She opened the door and glared into the darkness. She took another step, the door shut, and she leaned against the door from the outside. There was complete darkness. The wind had let down, but the air was heavy with dust and blocked out the starlight. She used to be out in the middle of the desert at night and never was afraid. But this was different. She felt vulnerable, exposed, and scared. Perhaps, the lack of light, she thought. She turned on the flashlight that was affixed on the left sleeve of her suit. She scanned the instruments around the void. It looked like they did not take any damage and appeared to be still in the same position as when they arrived.

Something attracted her to the void. She felt an urge to move closer, but was scared at the same time. Her fear ran deep. That was troublesome, she hated it, but there was nothing she could do about it. She slowly moved forward as she held her breath. There was nothing to see, nothing special. The urge to move closer became stronger and stronger. As did her fear. It occurred to her to let Ken know, but what should she tell him? That she was afraid, but that there was nothing out there?

She moved further in being one hand length away from the void. Pearls of sweat assembled on her forehead. What was drawing her? Her heart pounded. Her mind was fighting fear and urge at the same time. It felt as if something wanted to connect with her, but something so mystic, immense, and beyond anything she knew that it freaked her out. She simply stood there for a few seconds before she merged with the void.

Ken jumped up in the bunker as various energy discharges produced an impressive light display in the dark Martian sky. Shrieking alarms filled the thin Martian air. He ran to the door, opened it, and saw Tanya laying on the ground next to one of the scanners that she must had taken down with her as she fell.

*****

Kitahari sat sunken in her brown leather chair. It was incredible how much paper work was going over her desk. It seemed to be that every little detail needed her signature. Soon Earth would ask her what food she fed to her pet hamster—that means if she had one. She got up and looked out of the window and could make out a few tiny dots in the night sky. Fellow asteroids, she thought, somehow entangled in the gravitational tug play with Ceres. The ground station started to look impressive. Fifteen telescopes geared to spy on one of Saturn's moons, from where more than fourteen years ago an alien invasion fleet had been launched that nearly wiped out all life on Earth.

She felt marooned over here on this outpost furthest from Earth. People back on Earth never forgave her and Jack that they let the alien queen go back to Tethys II after they were all safely delivered on Mars. Well, they didn't know how much devastation the termkins—that's how people now referred to the termite-kind aliens—had caused on their home planet. But even if they had known, would they have decided differently? At least Jack and Carmen were on Mars and could build up their plantations and raise their daughter. But being on this desolate outpost was surely no enjoyment. The outpost looked enormous with its gigantic telescopes, but the living quarters that housed the crew of twenty-four were quite modest.

Suddenly the door opened. "Commander Alpinari, your visitor is here!"

"Send him in."

A huge guy stormed into the room, grabbed her, and shook her in the air with her feet dangling above the ground.

"How are you doing, Kitahari?"

"Nice to see you, too, Vladimir! I see you did not lose a bit of your strength. How have you been?"

Vladimir released Kitahari from his grip and sat down next to her. "Oh, well, you know me. I always find things to keep me busy."

"So, I heard. You are second-in-command for the expedition into the outer Solar System. Do you think that is wise? We did not have any problems with the aliens the last fourteen years and that could stir everything up again!"

"You might be right, but also consider that out of sight, doesn't mean out of mind. We want to check out how much of a danger the termkins still pose. And also how our new technology fares if it comes to a conflict."

Kitahari was silent for a moment, but then responded. "All what I can tell you is that there is not much going on at all, at least based on our spy telescopes. It is nearly as if they were not there at all. Of course, we can't look into that moon, but either there is basically no activity beyond that or they are damn good in hiding it.

"But the void is still there, right?"

"Yes, though its transmission style seems to have shifted somewhat."

"Well, then they should still be there. And why should they leave anyhow? One more reason for us to check it out."

You know that our logic does not apply. They are thinking—if we can call it like that—much differently than we do."

"Yes, yes," Vladimir responded, "but there are some commonalities, like protecting the offspring."

"Yes. Well, anyhow, we will give you the best remote sensing info possible, so that you will not be ambushed when you push forward into the outer Solar System! When will you leave?"

"Tomorrow 4 a.m., on the Morning Star, the command ship of the three vessels."

Kitahari got up, walked toward the nearby cupboard and took out two glasses and a bottle of vodka. She poured in the liquid up to the rim of the glass, gave one to Vladimir and said "Good luck, old fellow!"

Vladimir nodded and mumbled "We will need it!"

\*\*\*\*\*\*

Ken sprinted to Tanya as fast as he could. His flashlight did not give him much light and there was no way to know whether Tanya was alive. She did not move so much was clear. He pulled her into the spacecraft and then took off her astronaut's suite. She felt warm. He tapped her to wake her up.

Slowly she regained consciousness.

"What happened?" she asked.

"That's what I wanted to ask you?" he replied. He took a deep breath. "What did you do?"

Tanya did not answer. She still looked spaced out.

"Ken, Tanya, please come in, come in!"

"Oh, the base station, Ken here."

"What's going on there? Are you alright?" Tanner from the base station asked.

"Yes, I think so."

"Tanya's life signals went off for a few seconds and all the instruments that monitored the void went haywire." Daniel Perera entered the control room. "First question, is Tanya o.k.?"

Ken briefly summarized what happened. "I don't think she is fully there yet", he concluded.

"O.k., come back right away!"

"It's night and the storm is still out there. And what about all the instruments that are damaged?"

"We will take care of this later. This is an emergency situation. Get the spacecraft back as fast as you can. Over and out."

The station was buzzing with activity. The medical team assembled in the emergency room, and prepared as good as they could.

Daniel walked down the central hallway, when a door opened from the other side.

"Oh, good, Jack, you are here!"

"Tanner sent a brief report to me."

"When are they coming in?"

"Should only be ten more minutes."

"Any thoughts what might have happened? We did not have any incidents in the last seven years, since we resumed the exploration of the void."

"No, not really."

"Oh, here they come!"

Tanya was carried in on a stretcher and Ken trotted behind. They went to the emergency facilities. Tanya was staring with an absent gaze. The medical personnel scanned her and hooked her up to the monitoring equipment.

Jack stayed next to her, took her hand, and whispered. "Can you hear me?"

Tanya nodded.

Jack glanced over to the medics who confirmed that she seemed to be stable.

"What happened to me?" she asked timidly.

"It appears that you were dead for a few seconds. You had some kind of interaction with the void and that apparently triggered some kind of reactions, but we don't know what. How do you feel?"

"O.k., I guess. Somehow awestruck. This thing drew me in. The urge was so great even though I was terribly frightened."

"Do you think it wanted to communicate with you?"

"I have no idea. But if so, what? I don't have any recollection what happened? I was out!"

"Well, first you need to recover. The medics will make sure that everything is o.k. and then we can talk more later, o.k.?

She nodded.

Jack got up and turned toward Daniel. "What do you make out of this?" Daniel asked.

"I don't know. The most obvious would be some type of communication attempt," Jack responded.

"With a void? It is not alive!"

"But it is connected to life. Perhaps it is a conduit."

"But what would be the message then?"

"Maybe the message did not get through? Or, the message was fear?"

"Or it had something to do that she was dead for a few seconds and then alive again?"

"Possibly."

"That's all the facts we have. I will need to report that to Earth right away," Daniel stated.

"Yeah, I'll hang around here and see you in the morning."

*****

The engines started humming. Vladimir was walking to and fro in the command module of the star cruiser. The three ships around Ceres were waiting for the command to advance into the outer Solar System. He looked into space, the unknown that laid before him and a cold shower overcame him. He really didn't want to face those termkins again, but the Earth government was right. He fought them before and he needed to accompany the exploration team. After all, he had been inside Tethys and saw their nest.

He looked at the crew. All stared at him in excited anticipation. Even more so than at General McLean, who was leading the expedition. McLean led the build-up of Earth defenses and a defense fleet for the last ten years, but had really no experience with the aliens. Yes, they made tremendous advances in technology, but would that be enough for the high-tech termites? There hadn't been any contact between them and the humans since their fateful encounter. They remained in the outer Solar System, while the humans stayed within the inner Solar System with Ceres being the furthest outpost. It was like a silent agreement. Was it wise to challenge the status quo? The aliens had probably a good idea what the humans were up to, but we don't have a good idea what the aliens did in the last fourteen years, Vladimir thought. They live below the ground and we cannot probe there. We don't know more than what looks like an occasional ship in orbit or on transit between the Jovian and the Saturnian system.

Well, either way, it was time to go back. After all, the termkins were the invaders.

"Fire up the engines," General McLean said tensely.

The engines began to roar and the three ships were accelerating with their first target being the Jovian moons.

*****

"Oh, it's nice for you to come out this far", Jack said and smiled.

"How are the crops doing?" Tanya responded.

"Oh, it is always a challenge in the Martian soil. And then to get the illumination quite right, and the nutrient cocktail. But," and he stopped for a moment," that's not why you are here. How is your recovery going?"

"I'm physically alright. That's at least what the doctor's are saying."

"But?"

"I still feel very disturbed, shaken, to my core, and don't know what to make of it. You of all have the most experience dealing with the aliens."

"Oh, I don't think it has anything to do with the alien termites. The voids are something entirely different, a conduit to something much deeper, connected to the essence of life."

"Go on, I'm listening!"

"I think the voids are the source of life," he made a pause, "and death. That's why you are still alive. I don't think there is any other explanation."

"I wanted to ask you to come with me and fly to the void", Tanya suddenly bursted out.

Jack rubbed his forehead. "Is Daniel Perera o.k. with it?"

"Yes."

"Mmh, o.k., let me just wrap up some things. I'll be in the shuttle in about thirty minutes."

Tanya felt relieved. She knew that she had to face the void again to get rid of the fear she experienced. The incident occurred three weeks ago, but it felt like yesterday. There was no better person to accompany her than Jack. She liked that guy although he was her father's age. Perhaps, that's what made her feel comfortable with him—a father she never had. If someone could help her with that issue, it was him.

Twenty minutes later Jack arrived with his gear. He smiled to her "Ready to go." She simply nodded, ignited the engines, and steered the shuttle to the side canyon of Valles Marineris, where the void was located. Arriving at the void's location made her shudder. Only slowly, very slowly, was Tanya climbing out of the shuttle. Jack inspected the site. Perera's men had put the scanners back in place and adjusted. Nothing particularly noteworthy struck Jack. He had been here many times. When he saw how difficult it was for Tanya to get closer to the location, he gently took her hand and guided her closer.

"I wonder why this happened to you?" he asked her. "I have been here many times before and so have been many other people. But nothing like this ever happened. Is it something special about you?"

"I don't have a clue. It could simply be the timing."

"True."

They were inching closer to the void until they were exactly standing where Tanya had her previous experience. However, nothing happened—other than that it was quite obvious that Tanya was very frightened.

"I was hoping you would remember a bit more," Jack finally said.

"So was I", she replied,"but the only thing I recall is the fear, a tremendous fear of something seemingly dark and forbidden, drawing me in, along some kind of conduit, but no thoughts whatsoever."

Tanya paused and then she continued. "It was taking by breath away. No, more than that. It seemed to shut down everything in my body. It kept me pinned down and came closer and closer!"

Jack recalled the experience that he had, once, a long time ago, when he was commanding Deep Explorer. He did a meditation exercise, following the advice from an old friend and teacher, trying to connect to the void of the aliens. But somehow he landed somewhere else. A dark entity was hovering over him taking his breath away, for a few moments only, it appeared, but he awakened soaked in sweat. The entity radiated out fear, so much fear he had never experienced before. But he got himself together and led Deep Explorer into battle, and after many losses to victory. Tanya interrupted him from his thoughts. "Do you think that dark and forbidden feeling comes from the alien termites?"

"No, not at all. We were in close contact with the termite queen for a long time, on our trip from their home world to Mars. It was so alien, yes, but despite all this there was a basic understanding. Some common will to live. And common concern for our young."

Jack paused. "I often asked myself why they invaded our Solar System in the first place. Perhaps, they just expanded into new territory, like many insect species do on Earth. But maybe, they were running themselves from something."

"If they were only expanding into new territory, why did they take their own planet with them?"

"Yes, that's odd. Though you may have to be careful. They are not humans, and think and act very different from us."

"O.k., let's assume for a moment that they were fleeing from some adversary, then that adversary could possibly track them down and hunt them down, right? And the same may happen to us!"

A shudder went through Jack thinking that he might have to face the dark entity again and worse if it actually would come into the Solar System. "If that would be the case, we would be in a very bad situation, especially if that adversary is more powerful than the termkins—which you have to assume, because otherwise the termites would not be fleeing from it."

"And then the only thing that could save us would be to ally with the termkins, right?" Tanya concluded.

"If so, that would not be very likely to happen, since we are now invading their territory as we speak." And how would we communicate with them?"

"You did it before!" Tanya interrupted.

"That was very different. We were in close contact touching the queen."

An awkward silence occurred. Both knew what the logical conclusions were. If they were right, and that was a big if, a huge speculative leap admittingly, they would have to get in contact with the termkins before Earth's ships did, and further they would have to convince the termkins and the human government to communicate with each other rather than to engage in battle.

They trotted back to the shuttle. "What are we going to do?" Tanya asked. "I'm not sure," Jack answered, "let us talk to Daniel Perera first.

*****

Io appeared in yellowish hue gigantically on the main screen when the engines screeched to slow down. McLean purposely chose this moon of Jupiter for the braking maneuver. Io was definitely not the right environment for the termkins to have a colony. He didn't want to be surprised by a sudden attack when they were occupied with slowing down. Io was simply too warm. Even Europa, the next big moon over, they would not favor. Too much radiation. He remembered what one of the planetary science experts told him once: "If you land on the surface of Europa and take your helmet off, the radiation will kill you before you suffocate." The termkins were surely hardy, but that was even too much for them. And they wouldn't like the liquid interior of Europa either. However, Ganymede and Callisto were another story.

"All ships conducted successfully the brake maneuvers and are battle-ready", reported one of the lieutenants. Vladimir had been worried about this part. Not that the termkins were known to react and adapt quickly, but they would have been vulnerable. But now he began to be worried that this was too easy and could be a trap. McLean peaked over to Vladimir: "What do you think?"

Vladimir returned the eye contact with the general. Not that he liked him that much. He had something annoying. It was not that frequent twitching of his eyes, though he wondered how the general passed the psych test. It was not his trunk-like posture or his oversized nose, which did not seem to fit to the rest of the face. Perhaps it was that he was kind of the "favorite child" of the Earth Council. But then, that really could only be an advantage. So far, at least, Vladimir thought, he moves forward in a prudent and cautious way, and there is nothing to complain about that.

"I think we should just sit here a while and have our remote sensing equipment turned onto Ganymede and Callisto. Also, we probably want to get in touch with Commander Alpinari to make sure we did not overlook anything while we were occupied with slowing down."

"Good idea, check with the commander!"

"Alright"

## 7.2  Suleiman

"Jack, come in," Daniel Perera said and indicated for him to sit down in the chair in front of his desk. "I have to tell you some exciting new stuff, but you go first. After all you wanted to see me!"

Jack told Perera about the discussion he had with Tanya and then paused to give Daniel time to respond.

"Well, interesting thoughts. I never quite looked at it this way. So you think the voids are a threat?"

"No, not the voids themselves," Jack replied," how could they as they seem to be so intrinsically connected to life."

"And they assemble matter in the perfect arrangement after it has moved through," Daniel interrupted, "Something you found out, correct?"

"Yes, but maybe they are communicating a threat. A threat that displaced the termkins and that may be the reason why they left their place of origin in the first place."

"So what you want me to do?"

"I like to try to use your large array to communicate with them."

"Just based on the hunch you have?"

"If you want to call it like this, I presume you are right."

"O.k.", Daniel responded, "let me think about it. And, of course, I have to clear this with the Earth Council."

"I understand. So what was the news you wanted to talk to me about?"

"Some smart scientist on Earth, named Suleiman Alhaska, noticed some kind of lighting up of the voids when Tanya had her encounter with the void."

"You mean the Earth void lighted up?"

"Not only the Earth void. All voids as it appears!"

"That can't be. Even if there were some resonance effect, there would be huge time delays, because of the time the light needs to reach us."

"Ha", Daniel responded," and that is the most amazing thing of all. All the voids seem to be entangled on a quantum level. That means he measured this effect in real time. Yes, in real time," Daniel emphasized again, to have it sink into Jack's mind.

"My knowledge of quantum mechanics is limited, but aren't usually only two particles entangled with each other, no less, no more. And information cannot be transmitted easily. That's why it took so long to invent the QE chip, which ingeniously circumvents the problem," Jack replied skeptically.

"You are right with the two-particle entanglement, usually yes, but it seems that all the voids are entangled with each other!" That means if a thorough analysis is done of the effect, we know where all the voids are located, possibly even in the whole universe!"

"Wow," Jack rubbed his forehead.

"Of course, we had only one such event and the equipment that picked up the effect when Tanya connected with the void wasn't really tuned to it. So, they have incomplete information and really would want that we reproduce the effect again."

"Why don't they just use the void on Earth and make their tests there?" Jack said, with a tone of anger in his voice.

"You know the Earth void is off limits. First, because of all the religious implications, second people back home are way too freaked out to tinker with it. So, the Mars void will be the one where this will be tried. And Suleiman is already on his way to Mars to supervise this in person and also coordinate this with his colleagues back on Earth. The Earth council is really hoping that they can count on Tanya and you!"

"We were back at the void and nothing happened."

"Well, how should I bring it to you? You have to try harder! You are the most knowledgeable person on the voids and Tanya previously connected. You are the logical choice!"

Jack set back into his chair and breathed out a bit more heavily. "And I assume that if I agreed, they would consider my request to use the array trying to communicate with the termkins much more favorably, correct?"

Daniel Perera just smiled.

*****

Suleiman looked over his data again and again. What could he still figure out? The long duration of the flight to Mars was welcome news to him as he could in peace pour over the results obtained so far. Yes, the data would not lie and various patterns revealed themselves. First of all, there were many voids although not as many as some might have suspected. It was difficult to figure out how strong the voids were from the resonance effect of the quantum entanglement, but it appeared most were small. Fodder for the rare earth hypothesis, Suleiman thought, and evidence that most life in the universe is microbial. But what was the most surprising was that there were hardly any larger voids, meaning larger than Earth's void. Of course, it was a bit early to tell based on the data he could collect from the surprise event when that Martian astronaut caused the resonance of the voids. However, their absence was astonishing. Would that be the solution to the Fermi paradox, the reason why we were never contacted by intelligent extraterrestrial life before the encounter with the termkins? Suleiman's thoughts drifted off. How long is it anyhow until I reach Mars?" He checked the board computer. "2 weeks, 1 day, 14 hours, and 7 min!" The 38-year-old scientist of Arab descent, best of his class when graduating from MIT, was considered a loner and yes, he liked to be by himself, but even for him this was getting too long. Nevertheless, it made sense to be on Mars and analyze the void first hand. And it would also give him the opportunity to meet Jack Kenton in person. And to figure out with him together how to reveal the secrets of the void.

*****

"We scanned Ganymede for days. No evidence for any activity whatsoever," Lieutenant Baker reported to Vladimir. This is consistent with what I heard from Kitahari, Vladimir thought. He stared out of the big window of the Morning Star onto the giant ice moon Ganymede. Only ice and rocks, or so it appears, Vladimir thought. The other two ships were orbiting Callisto, but their scans didn't reveal anything interesting either.

"What are your recommendations, Dr. Kulik?" General McLean addressed him.

"Well, we have two options. We can still stay longer here and scan for possible termkin tunnels or subsurface habitats on those icy moons or we can proceed to the Saturnian system, and approach Tethys II. I suggest the latter."

"There is still no observed activity, not even from Tethys II?"

"No, nothing, sir. I just confirmed again with Commander Alpinari."

"O.k., I talked to the Earth Council and they confirmed with me that if we don't have any change in status to proceed with uppermost caution toward Saturn. So, you concur with this assessment?"

"Yes."

"Tell the other ships to assemble around us, and we'll fire up the engines tomorrow morning 10 a.m. standard time."

"Very well"

*****

Jack sat with Carmen and his 14-year-old daughter Marta at the dining table playing a board game that was intended to help Marta with her lacking geography knowledge of Earth. Marta was much more into the technology of the base station and the sites on Mars. She simply felt home at Mars and not at this blue planet in the sky that was so far away and seemed strange to her in many ways. Suddenly, a communication request popped in from Daniel Perera, and Jack went to the bedroom to take it.

Carmen followed him and asked "What is Daniel writing?"

"Oh, the Earth Council approved my request on trying to use the telescopes to communicate with the termkins."

"Wow, that is great. And what is the hook?"

"Well, I had to agree to try whatever I can to recreate what happened to Tanya. They want to study the resonance effect."

"But that nearly killed Tanya," Carmen protested.

"We don't know that."

"Just remember you have a daughter now, too. Two women you have to take care of!"

Jack turned around and moved his hand through Carmen's hair. "Yes, I do, and I will do whatever I can that nothing bad happens. But this is essential, it ties it all together, why are we here? What is our purpose in the universe? Life, the universe, everything!"

Carmen paused for a moment. "When is Suleiman arriving?"

"Ten days. And I will use the time to think about both—how to deal with the voids and how to communicate with the termkins. Perhaps, you want to help out with the communication?"

"Yeah, I think so," Carmen cautiously replied, "it looks like you are accepted back in the Ole Boy's Club."

Jack laughed. "Oh, I don't think so. But I do think they found a purpose for me again."

## 7.3 Doubts

The three vessels accelerated through space. Vladimir noticed that the nervousness of the crew increased steadily. They would have another six weeks to reach the Saturnian system. Many of the crew had experience fighting the termkins, some on the ground and some as part of the fleet that opposed them in space. And the helplessness of the Earth fleet stuck vividly in their memory. Sure, they made technical improvements, and the armory of these ships was hardly comparable to those that faced the termkins more than fourteen years ago, thought Vladimir, but it did not make him feel much better. They would be outnumbered by far and the technology of the termkins would still be superior, even if termkins had not advanced since their last encounter. Given that situation, was it really wise to break the untold truth? Surely, for the termkins, they represented now the invaders.

Vladimir turned from one side to the other. Thoughts rushed into his mind and then left again. It was critical to keep contact with Kitahari. Her giant telescopes on Ceres were their additional eyes. They definitely did not want to run into a trap and once they would slow down when approaching the Saturnian system, they would be most vulnerable. The aliens had shown that they were smart. Just kind of slow, which is the only really advantage we have, Vladimir thought. It felt good that Carmen joined the effort again. Well, at least in some way. The plan was that Carmen and Kitahari were trying to figure out a way how to communicate with the termkins. Of course, previous communication attempts had not been successful, but all those were attempted during the previous conflict, not after the termkin queen released them on Mars.

He would have also liked to have Jack more involved, but Jack only took care of the initial arrangements with the Earth Council and then handed it over to Carmen, since he had to focus on preparing the probing of the void with this guy coming from Earth. Yes, Suleiman, was his name. It was not clear to Vladimir how that could possibly help his mission, but it certainly was important. After the termkins retreated from the inner Solar System, the void research went onto the backburner. People back home were more concerned recovering from the onslaught of the termkins and build up defenses in case they would get another visit from the termkins. That took an immense amount of resources. And besides, no one wanted to mess with the Earth void. The void they could experiment on was the one on Mars, and that one became only now a priority again after the latest incident with that Martian astronaut.

Vladimir's thoughts drifted back to his mission. Maybe he shouldn't have volunteered for it? He had to admit to himself that he really didn't care to see those damned aliens and their planetoid again. But the mission needed someone who had been there before. Jack and Carmen would not go, and Kitahari not either. So, it was up to him to face the old enemy again. Six more weeks, six more weeks, it pounded in his head. Without him the mission would surely be doomed, but it was not clear if he would really be able to make a difference.

*****

"Suleiman, nice to meet you finally in person!" Jack greeted the arrivee.

"Thanks, it was surely a long trip."

The two men walked in the Atrium and Jack posted Suleiman on his latest preparations for the trip to the void.

Suleiman did not want to lose any time testing some of his new ideas how to activate the voids. Besides he felt that he wasted enough time sitting cramped up in this little shuttle that brought him here.

Nevertheless, Jack gave Suleiman a tour of the station, before showing him his resting quarters.

"See you tomorrow at sunrise!" Suleiman mumbled after viewing them. He had looked so much forward of finally arriving on Mars. But now, when he was finally there, he was dead-tired.

******

Carmen peered through the telescope. Tethys II appeared as a large, dark moon. "How the heck do you communicate with someone or something that uses some kind of telepathy?" she said to herself.

She went through her checklist on how many different approaches they tried previously during their first encounter. Light signals, radio signals, geometric shapes. What could they have missed? It was clear that communication was the key. Somewhere out there was Vladimir and the three Earth vessels, and the last thing anyone wants is a new war, she thought.

So she went through the list again and again.

She tried various algorithms, mathematical abstractions, geometric forms. But hell, they did try this all before. It had to be something what they did not try before.

But what could that be?

She contacted Kitahari on Ceres, but she had no new ideas either.

Heavy hearted Carmen went to bed that night. She needed an idea, one idea that would move her forward.

*****

"One week until orbital insertion around Saturn!" Lieutenant Baker reported to Vladimir.

Vladimir felt sick to his stomach. Was that really what he wanted to do, to meet those termkins again? Not really, but there was no turning back anymore. He went too far, they went too far. Vladimir trotted to McLean's quarters.

"Still no signs from the termkins?" General McLean greeted him.

"No, nothing. We are starting to prepare for orbital insertion."

"Great, that's wonderful. Or, don't you think so?" the general asked.

"Too good to be true, if you ask me," Vladimir responded.

"You would rather that we would have to fight ourselves all the way through to get here?"

"No, of course, not."

"But?"

"There is clearly something we don't understand. Where are they? Do they prepare a trap for us? Or, did they simply all vanish? And if so, why? I have no clue."

General McLean's forehead appeared more wrinkled than usual. "Well," he said, "you clearly have a point and I do trust your judgment. After all you were the only one of us out there, and also in close contact with one of their queens for months." McLean stopped for a moment, but then continued, "there are no interesting sightings from our friends on Ceres or Mars, either?"

Vladimir just shook his head and McLean continued.

"Well, either way we will know very soon, won't we?"

"I suppose so," and with this comment Vladimir left the general. He wandered through the corridors. The crew seemed to be in excellent mood. Was it really only him who was plagued by those dark thoughts?

\*\*\*\*\*\*

The shuttle landed with Jack, Suleiman, and Tanya on board just a few hundred meters next to the void. Suleiman was determined to trigger the void. Jack suggested a more cautious approach, while Tanya was already frightened just to be near the void. They geared up into their suits, opened the door, and walked over to the void. A few meters away they halted their approach.

"I have looked through all the images, video, and data, but it still amazes me that you can't see the void at all and there is no obvious way to know that it is even there," Suleiman said.

"Oh, yes, voids are real and there," Jack responded. "And this one is well marked, within a few millimeters", and he pointed to a red translucent sphere that was surrounding the void.

"Very well, let's get then the pulse laser out of the shuttle."

Tanya and the two men moved the tripod and the laser from the shuttle to the vicinity of the void and adjusted the laser to target the center of the void. They also set up various measurement devices, including a novel quantum state monitor to reveal any resonance effects, particularly with other voids, and linked those to the computer cluster on Mars. The adjusting and readjusting took them more than an hour.

"Well, we will soon see whether all my work over the last months paid off," Suleiman said with some clear excitement in his voice.

"Do we really want to do that?" Tanya restated some of her earlier opposition to the plan.

Jack remembered the disaster on Venus many years ago. Carmen was very nervous what they were about to do. But then they needed to make progress. It has now been close to 20 years since the discovery of the voids, he thought, yet we still don't know much about them, let alone their purpose.

It was clear that Suleiman considered Tanya's comment only as a rhetorical question as he prepared to fire off the laser. He also double-checked all the links to the computer cluster. Then he made a final call to Mars Base Station.

"Do we have a go?"

"All systems are working and we are ready. Whatever happens we will have it recorded and it is also sent directly to Earth," Daniel Perera responded. Carmen, who took a break from her communication duties stood next to him.

She felt quite unhappy about the whole experiment, but was at the same time quite curious what they would observe.

Suleiman, Jack, and Tanya moved behind a large boulder and Suleiman remotely activated the laser. Laser pulses were hitting the center of the void, which was radiating out a flux of constantly changing colors. Their intensity was increasing the longer the laser was shooting those pulsed frequencies at the void. As Suleiman was increasing pulse strength and time duration, the void was launching out tongues of spectral illuminations into the cold Martian air. Jack and Tanya stared amazed at the light show. It reminded him of fireworks on New Year's Eve in his native Switzerland. But Jack started to become weary and asked Suleiman "Are you absolutely sure that this can't destroy the void?"

"Yes, no worry." Suleiman responded. Then he called in to Perera: "Do you see any resonance effect with the other voids?"

"No, sorry, lots of energy spikes and interesting displays, but no resonance effect as the one we saw with Tanya," the Mars Base Station Commander responded.

"Damn," Suleiman responded, "something must not be adjusted right." In this moment he sprinted to the laser to adjust it, but at the moment he reached it, an energy tongue emanating from the void hit him and the laser at the same time. The resulting energy spike was so glaring that it blinded Jack and Tanya for a moment.

After their eyesight adjusted, the void was invisible again with no energy emanations at all. Suleiman laid next to the pulse laser on the ground with broken out glass from his helmet. Jack ran up to Suleiman, but could only confirm that he was dead.

"What the heck happened?" Daniel Perera shouted through the com line.

Jack reported what he had seen in a somber voice. At the end he added, "I guess, Suleiman wanted it so badly that he couldn't take it to see himself fail."

"Well, he did not," Perera noted, "he got the resonance effect he so badly wanted."

"But at what a price," Jack commented.

Carmen confirmed: "Yes, the quantum monitor showed for a second all the other voids resonating with the Mars void."

Jack was wondering what she meant with all voids, but figured that this was not the right moment to ask. He just relayed back to Perera that they would recover Suleiman's body and bring it back to base station.

## 7.4 Voids everywhere

Carmen was angry. What a senseless death. Was he so stupid to challenge the void that aggressively or did the Earth Council put him up to it? There was no reason to force the situation like that. Did he read the reports about her experience with the void on Venus? Carmen was confused, feeling at the same time sorry about Suleiman, but also angry at him. The best thing for her was to get back to work, get her mind off this and try to figure out how to communicate with the termkins.

So she went to her lab accompanied by Marta. School was over and it was late in the evening. Marta didn't care to stay by herself and Carmen thought she might be of some help. But Suleiman crept back into her mind. If he had only a brain. Oh, yes, he had a brain of a genius, but apparently his brain was not set up for common sense behavior.

"Why are you looking so distressed?" Marta asked.

"Oh, I'm just upset, because we have a brain, but don't use it the right way!"

"Do termkins have a brain, too?"

"Well, yes, tiny, but the queen has a much bigger one."

Wait a minute, the termkins communicate telepathically and the queens have a big brain and making the decisions, Carmen thought and ran to the other table. What did the scans reveal of the termkin queen when Jack and the rest of the crew flew back with her to Mars. That is the key, she thought. "We have to directly communicate with the brain waves of the termkins," she shouted to Marta, "come, help me, let's dig up all we can find!"

They started analyzing them. Patterns started to emerge. Each time Carmen got stuck, she called Kitahari for some background info. Eventually she got Kitahari hooked into the idea as well. Carmen and her daughter on Mars, and Kitahari on Ceres, worked via their communication link on deciphering the patterns and what they meant until late in the night. Some meanings became clear right away, others seemed to be impossible to decipher.

Eventually Carmen asked Kitahari "Can we transmit those patterns if we need to?"

"Yes, we have strong emitters, but it would be better if they were sent out from our vessels near Saturn. I'll put a tech crew on that right away to figure out how to do that!"

"Yes, please do so," Carmen replied, "our three vessels are about to enter orbit insertion. I don't feel good about the whole thing. We might need it quickly!"

Then, Carmen realized how late it was and an hour before sunset she went with Marta back to their living quarters to catch some sleep.

*****

The death of Suleiman was a somber reminder for everyone on the base station how dangerous it was to mess around with the voids. While members of Perera's men still worked on an autopsy of Suleiman's body, Jack stepped into Daniel's office.

"Just a tragic end of a genius of a scientist," Daniel greeted Jack.

Jack nodded. "He sacrificed himself for research."

"The sacrifice was not useless though. In the moment the energy field of the void got him, when he died, the resonance effect was triggered and all the voids lighted up. The voids are all quantum-entangled with each other."

"They lighted up?"

"Not literally, of course," Daniel continued, "but we saw their signatures in our quantum monitor."

Jack became curious. "So how many are there?"

"Literally millions of them, and at least thousands, no, ten-thousands in our galaxy alone."

"Why didn't we see this earlier?"

"Well, the same thing probably occurred when the accident happened on Venus with Carmen, but we did not anticipate anything and also did not have the modern quantum detection instruments we have today. Hell, we didn't even have the QE chip developed yet that nowadays is on every spacecraft."

Jack became a bit somber when Daniel mentioned Carmen's accident on Venus and recalled all the pain that Carmen experienced and all the torturing feelings he had. But then he got himself together again:

"That means life is very common in the universe."

"Yes, that is the interpretation. There is more though: our colleagues from Earth could relate the quantum signature to the size of the void, and relate it to the voids of our Solar System. The result was that more than 99% of the voids are really small, Mars-size, and there are only a very few that are Earth-size or close to it.

"That means life is common, but complex life or a complex biosphere is quite rare, consistent with the rare earth hypothesis?"

"Yes, it appears so. But the main odd thing is that they, at least so far, have not discovered a single void, that is more than 10% larger than the Earth void."

Jack was silent for a moment and then replied:

"Well, this would be consistent with the Fermi paradox and indicate that there is no alien intelligence further, or at least not much further developed than ours."

"That's what I thought as well," Daniel jumped in, "but if so what happens to voids and their civilizations that expand beyond ours? Do they just vanish or collapse? Given the whole universe and no voids larger than ours may provide a superficial explanation to the Fermi paradox, but it begs more questions than it answers."

"Isn't that always the case in science? But yes, and it has also direct consequences to Earth and our void. " Jack paused for a moment, was about to turn and leave, but then reconsidered and asked:

"Is there another large void close by?"

"Actually, yes," Daniel responded, "aside from the termkins void, there is one other one within 100 parsecs or so. It is located on Gliese 581d, a bit over 20 light years away.

"A super-earth planet around a red dwarf star, right?"

"Yes, in a Mars-type of orbit. But since it is a super-earth, it apparently retains enough atmosphere and liquid water to keep it warm and habitable. Something that has been suggested before. Our focus was directed on some of the other exoplanets, but, of course, now all telescopes will be geared onto the Gliese 581 system, especially its fourth planet. Based on the size of the void, there should be at least animal life, if not more."

"Quite intriguing. Carmen will be completely inspired about this!"

"Tell her my best wishes," Daniel shouted after him as Jack left his office.

*****

The three men sat nearly stoic on their seats as they entered orbit around Saturn. Vladimir on the right side of General McLean, and Lieutenant Baker on the left. Commanders Turris and Kenshaw from the two other vessels reported per com link that their orbital insertion was successful as well—as planned. Yes, as planned, Vladimir thought. The last time he was here he was hiding in the rings of that planet for months.

"Dr. Kulik, what do you suggest?" the general addressed him.

"Proceed slowly to Tethys II."

"So, it'll be."

The ships moved into formation with the Endeavor and Commander Turris on the right flank and the Spirit and Commander Kenshaw on the left flank of the lead ship.

Vladimir had difficulties to remain seated—the anxiety within him seemed to explode. They had various contingency plans, depending on what they encountered. But the truth was that nobody expected to encounter any resistance whatsoever. It felt that they were completely unprepared.

Vladimir opened the com link to Kitahari. The QE chip provided a blurred image of Kitahari, but her voice came through loud and clear.

"Nothing extraordinary that we can observe from here, Vladimir."

"Yeah, we don't see anything either. Simply no trace of the termkins. What do you make of it?"

"No idea. Wait a second. Let me initiate the three-person intercom. Carmen is joining us."

Carmen's image appeared, also a bit blurred, on Vladimir's monitor next to Kitahari's face.

"Oh, so nice to see you again. I guess, after all, you decided to join the party!"

"I assume Kitahari told you about our efforts with our new communication approach."

"Yes, the set-up is complete, everything is installed," Vladimir confirmed.

"O.k., then I'll be transmitting the instructions, frequencies, and pattern. Those should link directly into their brain waves. We just finished and that's what we got. No great data bank, but their communications seem to be pretty basic, not as complex as our human language. A lot of stuff we simply can't communicate, but again that's what we got."

Vladimir loaded down the transmission into the main computer terminal. It just took five seconds.

"Thanks a lot," Vladimir concluded," that may come in quite handy."

The small fleet of three vessels drew slowly, very slowly, closer to Tethys II. It seemed almost that they never wanted to arrive there. Each of the ships was in high alert, all weapons were charged and ready. But it would take another five hours to reach the target—assuming that nothing would happen on the way. But nothing did happen and so they drew closer and closer to Tethys.

Vladimir studied the communications modules. Yes, they were very rudimentary. But still they might become helpful, perhaps even critically helpful. It was ready to go from the main communication portal.

Finally, after a long, extremely long five hours they arrived at Tethys II. Vladimir could make out the opening where Jack's crew, including him, entered the subsurface labyrinth nearly 14 ½ years ago. General McLean ordered their engines to stop. They had arrived and were hovering over Tethys II.

## 7.5 Termkin jeopardy

"What's next?" General McLean asked his senior officers.

"The contingency plan asks for that we send in a reconnaissance group to enter the termkin's nest," Lieutenant Baker replied.

Even so that was the plan nobody seemed to be too excited about it anymore.

"Given that they are so concerned about their nest, that would definitely be considered an aggressive move or invasion," McLean replied.

"We crossed that line already long ago," the lieutenant replied and his doubts were renewed whether the general had it really in him to lead such a mission. Was he really the right man?

"I have another suggestion," Vladimir replied. Before we enter their sub-surface structures, we should make use of the new communication modules that we received from Commander Alpinari. There is really nothing to lose by trying it."

"We might lose the moment of surprise," the lieutenant countered.

Vladimir got annoyed with the lieutenant. "Look, I fought those aliens. They are not that dumb and there's lots of them. Our mission is simply too find out what the situation is and not to conquer back the outer Solar System."

"But if we can, that would be a major achievement," Lieutenant Baker interrupted him.

"With three ships? This would only be possible if they are already all dead and if so we don't lose anything by first trying to communicate with them."

"O.k., attempt communication then," said the general and ended the discussion.

"Very well," Vladimir replied. "This will be a bit improvised and we'll see how it goes." He conferred with the communication officer on the bridge and then said to the general "We will start with the brain wave pattern of *fellow worker*"

"Fellow worker?" the lieutenant asked dubiously. "Yes," Vladimir explained, 'they have a hierarchical state structure similar to our termites or bees on Earth. This should convey that we are fellow beings, colleagues, even friends, but not enemies or foes to be destroyed."

The lieutenant just shook his head when Vladimir mentioned the word "friends". The general didn't like it either and mumbled something to the extent that those friends killed millions of humans.

Vladimir ignored those signs of discontent and initialized the transmission.

They waited a minute, two minutes, nothing. Just as Vladimir was about to abandon the attempt, a brain wave pattern was received.

"It works, it works", the communication officer next to Vladimir declared. "Yeah," Vladimir shouted, "but the one million dollar question is know what it means! Reconnect with Commander Alpinari and the Mars Base Station. We need any support we can get to decipher this. This is absolutely critical."

Vladimir sweated profusely. He knew how important it was to interpret the termkins signal correctly. Their very life could depend on it. Meanwhile five minutes had passed and surely the aliens waited for a response.

"Carmen and Kitahari, our interpretation of the signal sent to us is "*No fellow worker, enemy!*" Do you concur?" Both women concurred. "So, what do we send them?" Vladimir appealed to them.

Carmen, who was meanwhile joined by Daniel Perera and a few others in the Mars Base Communication Center, replied "*Fellow worker, not enemy, all of us are life forms with young.*" Vladimir just looked briefly up to the general, who nodded, and then transmitted that message.

It took only a few minutes, but it seemed like eternity before the next brain wave pattern emerged. Right after they received the signal, a sphere of ships became visible that surrounded them.

"Damn, that are at least a hundred ships, and we are totally trapped", cried out the lieutenant.

"How the hell could that happen that we couldn't see them?" shouted the general. Through their panic, Carmen, who was linked in with Kitahari and Vladimir, and started right away the translation, came up with the translation of the received brain wave pattern: "*We only fought adults. You killed our unborn, destroyed our nest.*"

"How do we respond?" Vladimir shouted out. It became clear that they were at the mercy of the termkins. Not more than pins in a shooting alley. Even the general and lieutenant realized that. They were already eager to declare victory, yet the termkins must have used some technology, perhaps newly invented, that made them cloaked to visible light and their scanning frequencies. That response was their only chance. In some way it was surprising that the aliens did not blow them into bits already and still waited for a response.

The control rooms of the Mars Base Station and on Ceres were jammed with people now. Even Jack and Marta were in the communication hub. Kitahari took the initiative:

"Tell them something about when we kidnapped the queen but didn't kill it and let it return to Tethys II."

That seemed to be a good idea to Vladimir. What other options were there? He didn't even check with the general, but started assembling the wave pattern response. Time was of the essence and it was unclear how much patience the termkins would have. There was no pattern in the termkin's communication pool for the word "apology", perhaps it didn't even exist in their communication set, since it was based on intimate telepathy, so he drafted the following and sent it off:

"*Killing the young was a mistake, but we saved young when we sent the queen back to the nest with unborn young.*"

They waited for a minute, another minute. The tension was so high in the room that he could have cut it with a knife. At least they were still alive,

Vladimir thought. The question was for how long. The control room on Mars and Ceres were also completely quiet. Meanwhile, the Earth Council control room was linked in as well. The three ships were armed and battle-ready, but there was no question that their survival times would be no longer than a few seconds, if the surrounding ships were all firing on them. They were sitting ducks with the engines down.

Another five minutes passed by. No obvious action from the ships and no communication either. That was their language and they should be faster, Carmen thought, but then they might communicate with each other to come up with an agreed upon response. Or, was she thinking too much in human terms, and the queen, or lead-queen perhaps, would make the decision just on her own? There was no way to know.

The image looked surreal. Three earth ships hovering over a planetoid surrounding by a sphere formation of—as it became clear now—124 ships, with the three earth vessels in the center of the sphere.

Other questions popped into Carmen's mind. Would that mean a new war? Would they come and attack Ceres and then Mars with their cloaked ships? Would we be able to find some countermeasure to make them visible again? Would Earth send a whole fleet to fight the termkins if they destroyed the ships?

Another few minutes passed. "Did they receive the message?" General McLean asked impatiently. "Yes," Vladimir replied, "we are repeating it all the time until we get the next communication from them. Then Vladimir had another idea. He assembled another communication and sent it off:

*"We have unborn young with us. We want to go back to our nest!"*
"Why did you send that?" the lieutenant approached Vladimir.

It occurred to me that some of our women might be pregnant and we can be considered in a similar situation as the termkin queen, when Jack Kenton let her go back home from Mars. We showed mercy, compassion, or however you want to call it. Perhaps they return the favor to us.

Some more minutes passed by. Then a message came in from them:

*"Go home to your nest!"*

Vladimir's anxiety-worn voice shattered through the com link: "Carmen, Kitahari, do you concur with my translation?" Both women gave their affirmative.

Some of the termkin ships started moving.

"Don't shoot, don't do anything hostile," Vladimir shouted from the communication console to the command center and to the other two ships. "Let them finish whatever they do, they are acting slow." The sphere of starships transformed into a funnel with an opening directed toward Earth.

"Engage engines, and let's go home!" the general commanded his ships. The ships started accelerating on their way home and were soon out of the reach of the termkin ships and their home moon.

## 7.6  Speculations

Six months later, Mars Base Station, Conference Room.

Jack and Carmen waited already in the conference room with Daniel Perera, when Vladimir entered the room. Carmen, who didn't see him when he arrived a few hours prior, was the most emotional and Vladimir gave her a big bear hug. After the beginning courtesies, Daniel established the com link to the representative from the Earth Council, who joined them, and also to Kitahari on Ceres. After the com links were shown to work fine, he started:

"Welcome to our meeting. We are here to discuss some of the implications of the latest events. After allowing Dr. Kulik to attend the meeting by providing him with a shuttle, General McLean is on his way to Earth. However, we do have all the reports from him, and Dr. Kulik can correct us, if needed. Let me give the word to our representative from the Earth Council first, Mr. Kelley?"

"Thank you," a man with a grayish mustache and a stern look responded. "The Earth Council is mostly interested what your assessment is how serious and imminent a danger the termkins pose. Also, if there is any connection with the voids, and what they represent. You have been all in close contact with the termkins and voids, thus I'm looking for your insights, and then represent the case to the Executive Council, which might take actions from a pool of various options.

Vladimir answered first: "I do not believe that the termkins are an immediate threat. We had peace for more than fourteen years and now were able to establish communication. They could have destroyed us, but instead let us go."

"So why did they let you go?" Mr. Kelley inquired.

"I think because we succeeded in a common understanding. On one point only really, the protection of the young, which gave us a common goal."

"There may be more to it," Jack interjected, "when Suleiman triggered the quantum resonance effect in the voids, their void lighted up as well. They might have observed the same phenomenon we did. All the other voids, and especially theirs and ours, another common thread."

"This is highly speculative though," Kitahari added, "as we do not even know whether they can observe the quantum resonance with the instruments they have."

"True," Jack responded. "However, we know from our first encounter that they knew what the voids are, or at least how to destroy them. They did so with Titan's void."

Carmen really did not like the reminder of that fact. "Despite," she added, "Titan having a void, meaning common ground, they destroyed it."

"That maybe the case," Daniel interjected, "but Earth's void is much bigger, so they might have considered us more as equals, especially since they didn't win the previous war."

The man representing the Earth Council said, "What danger do the termkins pose today?"

"I think as long as we stay in the inner Solar System, they leave us in peace," Vladimir thought out loud, "but it would be very comforting if we can modify our scanners so they are not invisible to us."

"We are working on that," Mr. Kelley said, "and let me assure you that we are making good progress. That won't happen again."

"But they might come up with some other trick to surprise us, right?" Daniel commented, "after all they seem to be still ahead of us technologically."

"They are," Carmen said, "but we are developing faster, which means that at some point we are ahead."

"So, wouldn't that bear the danger that they would want to counteract that development with a preemptive strike," Mr. Kelley interrupted.

"We have to remember that we are not dealing with humans," Carmen continued. "Those aliens, the termkins, use more the strategy of our social insects. Their primary concern is their nest and the young. That's why they retreated when we attacked it and that's what I think the common understanding is based on. Their initial attack, now 15 years ago, was to expand the habitat and destroy all possible competition for resources. That did not work, so they are now content where they are and what they have."

"Does the rest of you concur?" Mr. Kelley asked.

"Yes, I also think that the termkins don't pose an immediate threat as long as we stay out of the outer Solar System," Jack responded. The others also gave their affirmative.

"O.k., then to the voids, what do you make of them?" Mr. Kelley inquired.

Jack made the first attempt to address this one: "They are some kind of quantum phenomenon associated with life, wherever there is life, there is a void, and vice versa. And the larger the void, the more complex life is."

"What do you mean with "complex life"? Daniel questioned.

"My hypothesis is that it has something to do with consciousness. Macroscopic and multicellular life is more aware and conscious than bacteria, and we have a higher degree of consciousness than animals. I don't think that it was a coincidence that each time the voids showed that quantum resonance was when human consciousness intersected with a void, once with Tanya, the other time with Suleiman."

"Of course, this hypothesis will be difficult to verify," Mr. Kelley responded. "How can it be that if you move some material through the voids that it gets perfectly replaced? And that they are all connected by quantum entanglement?"

There was a minute of silence in the room, then Vladimir addressed the question, "I thought about it a lot, but can't make really any sense of it. And probably no one else either. It seems to me that what the nerve cells are in our body, the voids are for the universe."

"Why are they here then? What are you saying?" Daniel commented, and Vladimir remembered the conversations long ago, which he had with Jack on the Venus Orbital Station, "Maybe there is no why, they are simply there as quantum manifestation of life, or perhaps the universe itself is alive in some kind of weird way!"

"What do you make of the apparent upper size limit of the void and the recent observations by Cussou?" Mr. Kelley asked.

"What observations?" Daniel responded.

"Oh, I thought everyone received the report. The results are still preliminary and open to alternative interpretations, but the leading hypothesis is that the energy pulse measured with the vanishing of a void, located in the Virgo Supercluster, was associated with the creation of a new universe, basically the formation of a new bubble or universe within our universe. And that this void was one of the largest voids that we had measured."

"Oh," Daniel said, "I didn't hear about that yet. What do you make out of it, Jack?" and he turned to him.

"It might mean that if life develops on a planet to a high degree of consciousness that this sparks the creation of a new universe."

"And that would explain the Fermi paradox!" Daniel exclaimed.

"And this would mean we live in a multiverse," Kitahari added.

"More than that," Carmen interjected into the discussion, "that may sound way off. But from a biologist's viewpoint that looks to me that the universe itself might be alive, even conscious, and each time consciousness spikes in one location of its space-time continuum, then a new baby universe is born with a set of new universal parameters. Of course, many of those would be uninhabitable places, but some would have the right mix of the universal constants, expand, and may spark even a new succession of universes."

"So, that would happen to us as well?" Daniel asked.

"Presumably."

"What interests me from a more practical application is," Mr. Kelley said, "assuming of course you are correct, which is a big if, when would that happen to Earth. Are we in any immediate danger?"

"I would not think so," Jack continued the discussion, "the termkins are technologically more advanced and their void is still there. Arguably they might have a lesser level of consciousness, at least on the individual level, but there are also voids out that are still slightly, but significantly larger than Earth's void."

"But there is really no way to know when the process would start," Carmen added, "and there is not really anything we can do about it."

"Well," Mr. Kelley responded, "I have all the input I need for the Earth Council executive meeting and will report to them on the practical implication that you don't consider the termkins and the voids to be an immediate threat. Everything else I consider wild speculations. Thank you for your time."

Mr. Kelley disconnected from the com link, as did Kitahari. Daniel Perera took over the initiative and said: "Why don't we continue the discussion over some special Martian beverages in our newly remodeled cantina. After all, we have to celebrate Vladimir's safe return—and of course, the safe return of all our ships. I don't think we will solve all the workings of the universe, why there is life, or even why we are here on this darn red barren planet—but at least we can enjoy the moment."

And Carmen added, "at least as long as it lasts. I have the feeling tomorrow we'll wake up in a different universe!"

# Part II

The Science Behind the Fiction

# Astrobiology—a melting pot of open scientific questions

## 1 Venus Orbital Station

Vladimir and Jack are monitoring microbial densities in the lower cloud layer of Venus. The possible presence of microbes in the lower atmosphere of Venus has previously been postulated by various authors [1–5]. The underlying idea is that Venus was a planet positioned in the habitable zone of our Sun—many billions of years ago—meaning that liquid water was stable on its surface. The earliest traces of life we know from Earth are at least 3.8 billion years old [6], possibly older [7]. Some researchers [2] propose that oceans could have existed on Venus for well in excess of one billion years. The argument then is: given that early conditions on Venus were very similar to Earth, and we know Earth had life at least 3.8 billion years ago, then life could have originated on Venus. Alternatively, it is well known that microbes can survive space travel being ejected by an asteroid impact from one terrestrial planet of the inner Solar System and deposited onto the other [8]. Thus, even if life did not originate independently on Venus, it could have been transferred from Earth at a time when liquid water was still stable on its surface. Once microbial life got established on the surface of Venus, it would be tenacious and trying to survive any ensuing environmental changes. Indeed, Venus' environmental changes during its planetary history were quite challenging as it became the greenhouse planet it is today with surface temperatures well above 450 °C with little water around. However, microbial life could have survived these shifts in environmental conditions if they occurred gradually enough and if they would have had an environment on Venus to retreat to. The surface is too hot (and dry) as is the subsurface, but the lower atmosphere is what many researchers would consider a borderline habitat for microbial life. This is based on the following observations [5] (1) The clouds of Venus are much larger, more continuous, and stable than the clouds on Earth; thus providing a more amenable habitat to life than Earth's atmosphere; (2) the atmosphere is in chemical disequilibrium, with hydrogen gas and oxygen gas, and hydrogen sulfide and sulfur dioxide coexisting (disequilibrium conditions in the atmosphere are usually considered a possible biomarker for life, like the disequilibrium of oxygen and methane is in Earth's atmosphere); (3) the lower cloud layer contains nonspherical particles comparable in size to microbes on Earth; (4) temperatures of 30–80 °C, approx. pH 0, and a pressure of about 1 bar are environmental conditions in the lower cloud layer of Venus that are tolerated by some microbes on Earth; (5) the super-rotation of the atmosphere enhances the potential for photosynthetic reactions by reducing the time duration without light (a day–night cycle of 4–6 Earth days compared with 117 Earth days on the surface); (6) an unknown absorber of ultraviolet (UV) energy has been detected in the Venusian atmosphere, which could be related to microorganisms that use a fluorescent material (e.g., cycloocta sulfur, the stable yellow sulfur you can buy in your drugstore) to convert UV-radiation into visible light suitable for photosynthesis; and (7) while water is scarce on Venus, water vapor concentrations can reach near 1 % in the lower cloud layer. Traditionally planetary atmospheres have not been considered as a likely habitat for permanent habitation. However, some more recent work calls this into question. For example, some scientific investigators [9] showed that bacteria in cloud droplets at high altitudes on Earth are actively growing and reproducing, and concluded that the limiting step for the persistence of microbial life in cloud droplets is residence time in the atmosphere. It turns out that particles in the Venusian atmosphere have much longer residence times than in Earth's atmosphere, on the order of months compared with days on Earth [10, 11]. Microbial reproduction can occur within hours, more commonly within days in the natural environments, but this still lies within the particle residence time of these droplets. Even if these cloud droplets sink and evaporate on average after a time period of one month, several cycles of reproduction could have already occurred. A space mission should be designed to analyze the cloud droplets for their composition to investigate whether these particles are microbes that are covered by an inorganic fluorescent layer. The Pioneer mission in the 1980s had as one objective to investigate the composition of these so-called mode 3 particles, but unfortunately, the instrument designed to measure it, a pyrolyzer, failed to function. Thus, until the day of this writing it is still an open question whether microbial life might exist in the lower cloud layer of Venus as a remnant of an earlier biosphere that started in the Venusian oceans.

# 2  Earth

Rigorous physical and psychological training is a well-known component which NASA uses to get astronauts ready for space missions. Especially the psychological stability under stress is a critical parameter [12]. The witch cage and some of the other methods used as described in the novel are known by the author from various martial arts training programs and could thus plausibly be employed for such a longer and extremely challenging space mission.

The quantum entanglement (QE) chip developed by human scientists on Earth during the time span of the plot is a fictional device that allows instantaneous communication over long distances based on the fundamental quantum mechanical phenomenon of entanglement. As an example, consider two photons (quanta of light) created jointly by the same quantum-mechanical process, having thus their physical parameters correlated. Prior to any actual measurement their state (e.g., polarization) is undetermined. However, once a polarization measurement is made on any of the two photons, the states of polarization of both photons are known simultaneously, however large the distance between them, since they have been created together in the first place. The current understanding of the phenomenon is that no information can be transferred from one particle to the other, though some recent research seems to undermine that notion [13]. Information transfer which is faster than light involves extraordinary problems with causality [14] in a relativistic universe.

# 3   Mars

A human station on Mars has been in the planning stage for many years based on NASA long-range objectives, but nowadays also by goals advocated by private space companies such as SpaceX. The idea that Mars is inhabited by microbial life is under intense discussion. On one hand, the current surface conditions make it for life (as we know it from Earth) quite challenging to survive in this nutrient-deprived, cold, dry, and radiation-intense environment. Organisms would not only need to survive for a limited amount of time, but also would need to metabolize and undergo reproduction. The only-ever-conducted life detection experiment (the Viking landers in the 1970s)—as of this writing—came back inconclusive. All three experiments conducted by the Viking landers observed chemical changes that indicated the possible presence of life, although the expected signals were not as large as expected for a biological response and tapered off with time, casting doubt on a biological explanation. This led at the time to the consensus view that Viking had detected reactive, oxidizing surface chemistry, but not biochemical metabolic processes [15]. However, Gil Levin, the Labeled Release Experiment's principle investigator maintains until today that his experiment proved metabolic activity by Mars microorganisms [16]. Joop Houtkooper and Dirk Schulze-Makuch proposed that Martian microbes might utilize a hydrogen peroxide–water mixture as their internal solvent (compared to salty water that is used by Earth organisms) as special adaptation mechanism to thrive at the near-surface of Mars [17]. This postulated adaptation would have several advantages as the solvent would stay liquid down to temperatures of -56 °C, it is hygroscopic (meaning it is like honey or many salts that can attract water directly from the atmosphere, which would be a great evolutionary advantage when living in a very dry environment), and it could be used in metabolic pathways as a source of oxygen. If Martian microbes had indeed a hydrogen peroxide water solution in their cells, it would also explain why little to no organics were detected with the Viking's gas chromatograph mass spectrometer (GC-MS) instrument, because hydrogen peroxide becomes unstable when heated and would oxidize all organics to carbon dioxide – and indeed a flux of carbon dioxide was measured in the GC-MS. The GC-MS, however, measured a trace of chlorinated organic compounds that was at that time interpreted as contamination from Earth. Rafael Navarro-Gonzalez and coauthors interpreted that signal as organic material from Mars, related to the perchlorates that were discovered by NASA's Phoenix Mission and also by the Curiosity Rover [18]. Perchlorates in the Viking samples when heated to 500 °C could have decomposed into reactive oxygen and chlorine oxidizing any organics present and producing the trace chlorinated organic compounds detected [19].

In the novel the discovery of the exploration teams on Mars is based on two analog environments that are studied on Earth. Mars has large salt deposits, especially sulfur salts. The place where this can be observed on Earth is Lake Lucero, which is part of White Sands National Monument. Here one can find large crystals of gypsum up to some tens of centimeters and enormous amounts of other salts, in an environment that is referred in geologic terms as a playa (a lake that is usually dry) [20]. The microbialites (mounds of calcareous material built by microbes) can be found in Pavilion Lake, Canada. These structures are up to a meter tall or taller and occur in a nutrient-poor mid-latitude lake, and must have formed in the lake sometime after the last ice age about 10,000 years ago. Yet, that distinctive assemblage of freshwater calcite (which has some similarities with coral reefs in seawater, but is not as diverse and only microbial in nature) at Pavilion Lake has been associated with organisms just before the Cambrian explosion about 550 million years ago [21]. The Cambrian explosion is referred to the time as Earth's biota emerged from the melting of the near-global ice coverage when organisms diversified in an unprecedented matter of new forms and building plans. Research at Pavilion Lake has confirmed that the organisms that build up consist entirely of microbes and communicate with quorum sensing molecules [22]. These organisms might also indicate a precursor community just before the ascent of multicellular life. The microbial community also includes cyanobacteria (with the metabolic strategy of photosynthesis). The fraction of cyanobacteria declines sharply in the deeper microbialite structures, thus representing still a possible analog to the Martian cave environment, where the fraction of the photosynthesizing microorganisms would be expected to be zero or close to zero.

# 4    Titan

Saturn's moon Titan is truly exotic. It has a substantial atmosphere (1.5 bar near the surface), with nitrogen and methane being the major compounds (98.5% and 1.5%, respectively), with methane concentrations approaching 5% near Titan's surface [23]. Other organic compounds are present as minor components. Measurements indicated the presence of methane rain on Titan [24, 25] and liquid methane-ethane lakes on Titan's surface [26]. The landscape may resemble an oil spill area in northern Alaska or Antarctica, only that it is much much colder. Yet, despite the frigid temperature at Titan's surface, 95 K (less than -150 °C), it has been speculated that possible life on Titan would use metabolic pathways that involve reactions with photochemical acetylene [27], hydrogen, and heavier hydrocarbons [28], all compounds that are present in Titan's atmosphere. Even a National Academy of Sciences report [29] concluded that the environment of Titan meets the absolute requirements for life. Titan's current atmosphere is likely to be quite similar to Earth's early atmosphere, about four billion years ago, before cyanobacteria developed that increased the atmosphere's oxygen content of Earth significantly.

There is not much liquid water to be found at Titan, probably only as an ammonia-water mixture deep in the subsurface, and possibly in a form of a deep subsurface ocean [30]. However, a hydrocarbon solvent may actually increase the chances for the origin of life. For example, extensive experience with organic synthesis reactions has shown that the presence of water greatly diminishes the chance of constructing nucleic acids [31]. And organic reactivity in hydrocarbon solvents is at least as versatile as in water, and many enzymes derived from organisms on Earth are thought to catalyze reactions by having an active site that is hydrophobic (nonpolar) [32]. Therefore, the construction of organic macromolecules in a hydrocarbon solvent may actually be more straightforward than in a polar solvent such as water or ammonia. Life though, if it were to exist in such an environment, would have to be quite different. In a hydrophobic solvent, cellular membranes would have to be constructed differently. In Earth organisms the membranes are hydrophilic (polar) to the outside to optimally interact with the polar compound water, and hydrophobic to the inside. Since the outside solvent would be a hydrophobic hydrocarbon compound on Titan, the structure would likely to have been reversed. Also, life with a water solvent has to be quite small, or at least the basic unit (the cell), because of surface-to-volume constraints. In a hydrocarbon solvent those constraints are more lose, and the basic unit could be much larger to overcome the low solubility of organics in liquid methane and enzymes to catalyze reactions at the low temperature [19]. Life expectancy is another factor which is very variable in principle. Our life expectancy on Earth is quite short, because we live in a relatively high-temperature environment (processes move faster with higher temperature) with high oxidative stresses (our macromolecules such as DNA have to be constantly repaired). In a cold environment such as Titan, processes would move much slower and there is little solar radiation, thus in principle the life span of an organism could be thousands, if not millions of years. Thus, the hypothetical Titan slugs are feasible as organisms in such an environment.

## 5   Deep Encounter

Many science fiction movies have been made about a first encounter with an alien intelligence. Usually these encounters are portrayed to end up in a disaster. There is a reason why. Any intelligent extraterrestrial species we would encounter in space would likely have a predatory instinct. Why is that? It is related to the reason that sheep are not very intelligent. The primary function of sheep is to graze and to escape when a predator is coming. The task of the predator is much more challenging, because it has to outsmart the sheep, catch it, and consume it. This requires anticipatory thinking of how the sheep will react once it recognizes the predator. Also, the predator has to align the movement of the sheep (presumably running away) with its own movement when pursuing the prey in a high-risk effort. The predator has to jump onto the prey, subdue it, and finally kill it. The predator has also to weigh in whether the possibility of injury, which in the wild means often death, is worthwhile the risk. Thus, it looks for a weak or very old or young individual. If the predator can hunt in a group such as in a pack of wolves, the predator is likely more successful. However, this will even require more coordination and intelligence, and also social interactions. Our ancestors, who brought down mammoths which are much bigger and heavier than an individual human, excelled at this endeavor. Thus, any extraterrestrial intelligence encountered would be expected to be a social organism with at least some predatory genes or instincts.

Even if the intelligent alien species were not to be hostile per se, a conflict could very quickly arise, because of a misunderstanding, or a rightfully or not perceived threat by the alien from us in order to protect themselves. Communication will be a major challenge. A good example of this kind of challenge is that we still don't know how to communicate with dolphins, which are no doubt an intelligent species and are even quite closely related to us. Encounters with dolphins are usually a positive experience for humans and there seems to be some mutual understanding, much based on the altruistic behavior of dolphins. But how would such an encounter play out in space, behind the walls of a spacecraft? And what kind of communication would we be able to establish with meta-intelligences (intelligence of the group or state), such as an ant, bee, or termite colony. They do not have the individuality as we do (or the dolphin). How could we even relate to them? The same is true on the other side of the individuality spectrum for an octopus. Although all of these organisms are related to us, we usually cannot communicate with them, especially not from a distance. The chance of successful communication is greater if they are related to us or at least match us in their behavioral pattern. However, even if we were to encounter a species with similar behavioral patterns, there would be plenty of opportunity for misunderstandings that could easily result in conflict. If and when we encounter an alien intelligence, chances are that they are not at all related to us. One last point to be made is that colonization attempts by humans of new lands, where indigenous peoples lived, usually didn't go very well for the indigenous people, who were at a lower level of technology. And these unfortunate encounters occurred between groups of the same species!

# 6   Tethys II

The aliens are based on some of the same physical characteristics and social structure as termites. Since termites (like other social insects such as bees and ants) have a highly sophisticated nest and breeding structure, even with air-conditioning in the case of termites, a more intelligent queen in that structure would logically be mostly concerned with the offspring, because that is the main purpose it serves. Social insects as we know them from Earth are not considered intelligent in a conventional sense, and as individuals they certainly do not meet the criteria usually used for intelligence. However, as an aggregate or group, they display some of the features that would suggest intelligence were they a single organism [31]. For example, they build elaborate housing, divide labor, communicate symbolically (at least in the case of the bees), radically modify their environment, grow and cultivate food (in the case of fungus-growing ants), domesticate other species, wage war, and cooperate for the good of the entire group [33–35]. As such, they represent a case of meta-intelligence [36]. The success of the social insect is testimony of the adaptive utility of this form of intelligence and that it should not be underestimated.

The alien termites in the novel live underground in a cold environment, thus they use an ammonia-water mixture as an internal solvent. Ammonia is a solvent very similar to water and in mixtures with water conveys amazing antifreeze properties. Also, since the alien termites or termkins live in this kind of environment where chemical processes proceed slowly, they also move rather slowly. Otherwise, however, they are more technologically advanced and the faster movement and thinking is the only advantage the human species has in the conflict, and the advantage is much less pronounced when you can't run away (e.g., when Earth itself is attacked). As a technologically advanced species it makes sense that they would take their home, i.e., their home planetoid, with them. As they are adapted to live in a cold environment, the heat of the planetoid is provided mostly from the radioactive decay of the interior rather than sunlight.

# 7 The Voids

The existence of the voids is fiction. However, there is in Buddhist spirituality the goal to reach the void, which refers to the absence of inherent existence in all phenomena, and it is complementary to the Buddhist concept of nonself. There is also the concept of vitalism, the notion that living organisms are fundamentally different from nonliving entities because they contain some nonphysical element or are governed by different principles than are inanimate things [37]. In modern science, especially with the discovery of DNA, vitalism has been replaced by more empirically mechanistic models. Vitalism explicitly invokes a vital principle, for example, why is one body alive while another is not, and that element is often referred to as the "vital spark", "energy", or in a more spiritual connotation is also equated in humans with the soul. There is also some thought that the reality we perceive is only a small part of what really exists, particularly when it relates to quantum processes. Thus, the voids are fictional entities woven together from these various concepts. In particular, the novel pictures all voids connected at the quantum level, perhaps being created and entangled when life (the definition and reason for it not further explained here) came into existence, which was intrinsically connected to the universe and related to a time when quantum fluctuations were dominating the universe's expansion (quantum cosmology). Exciting one of the voids—as done unintentionally by Tanya or intentionally by Suleiman—can thus be seen as quantum mechanical measurement of their state, providing simultaneous information about the other voids in the universe.

The discovery of the voids also touches on the Fermi paradox, which asks the question why we haven't been in contact with extraterrestrial intelligence given that there are so many stars and planets [38]. Surely, on some planets where life originated it should have developed to technologically advanced life forms? Usually the Drake equation is cited in this context [39], which—in its conventional form—states:

$$N = R^* \cdot f_p \cdot n_e \cdot f_l \cdot f_i \cdot f_c \cdot L$$

where:
$N$ = the number of civilizations in the galaxy with which communication might be possible
$R^*$ = the average number of star formation per year in the galaxy
$f_p$ = the fraction of those stars that have planets
$n_e$ = the average number of planets that can potentially support life per star that has planets
$f_l$ = the fraction of planets that could support life that actually develop life at some point
$f_i$ = the fraction of planets with life that go on to develop intelligent life (civilizations)
$f_c$ = the fraction of civilizations that develop a technology that releases detectable signs of their existence into space
$L$ = the length of time for which such civilizations release detectable signals into space

The Fermi paradox becomes even more puzzling, because there should exist many civilizations before us (and after us), which should be able to contact us using reasonable estimates. Also, many scientists conceived so-called von Neumann machines, a self-replicating machine that is capable of autonomously manufacturing a copy of itself using raw materials taken from its environment (after an earlier concept from von Neumann [40]). If there were only one such civilization in the whole galaxy and travelling by 10 % of light velocity, which is even for us realistic to achieve on long space flights (e.g., see, for example, the 100 year Starship Symposium held in 2011, web link http://100yss.org/symposium/2011), then the whole galaxy could be occupied by these autonomic machines in about four million years. Thus, the question remains what the solution of the Fermi paradox is. Many different hypotheses have been offered from the Zoo hypothesis (humans are kind of a wildlife preserve or, if you prefer, aliens may apply Star Trek's prime directive) to the idea that we live in a computer simulation [41]. Of course, the solution could be as mundane as that the origin of life is so difficult to achieve that there are not many life forms elsewhere in the universe (just making one factor of the Drake equation very small would make the number of civilization in our galaxy very small) and that any civilizations would be so far apart in both time and space, that a contact between them would statistically be extremely unlikely. In this novel, the solution of the Fermi paradox is provided by appealing again to the symbiosis between the physical universe and the phenomenon of life: once a void reaches a critical size (e.g., reflects a civilization that has become sufficiently large and intelligent) it vanishes by creating a new baby universe with its own development and destiny.

# References

1. Sagan, C. (1961). The planet Venus. *Science, 133*, 849–858.
2. Grinspoon, D. H. (1997). *Venus revealed: A new look below the clouds of our mysterious twin planet.* Cambridge: Perseus Publishing.
3. Cockell, C. S. (1999). Life on Venus. *Planetary and Space Science, 47*, 1487–1501.
4. Schulze-Makuch, D., & Irwin, L. N. (2002). Reassessing the possibility of life on Venus: Proposal for an astrobiology mission. *Astrobiology, 2*, 197–202.
5. Schulze-Makuch, D., Grinspoon, D. H., Abbas, O., Irwin, L.N., & Bullock, M. (2004). A sulfur-based UV adaptation strategy for putative phototrophic life in the Venusian atmosphere. *Astrobiology 4*, 11–18.
6. Schidlowski, M. A. (1988). A 3,800-million-year isotopic record of life from carbon in sedimentary rocks. *Nature, 333*, 313–318.
7. Mojzsis, S. J., Arrhenius, G., McKeegan, K. D., Harrison, T. M., Nutman, A. P., & Friend, C. R. L. (1996). Evidence for life on Earth before 3,800 million years ago. *Nature, 384*, 55–59.
8. Mileikowsky, C., Cucinotta, F. A., Wilson, J. W., Gladman, B., Horneck, G., Lindegren, L., Melosh, J., Rickman, H., Valtonen, M., & Zheng, J. Q. (2000). Natural transfer of viable microbes in space: 1. from Mars to Earth and Earth to Mars. *Icarus, 145*, 391–427.
9. Sattler, B., Puxbaum, H., & Psenner, R. (2001). Bacterial growth in supercooled cloud droplets. *Geophysical Research Letters, 28*, 239–242.
10. Esposito, L. W., Knollenberg, R. G., Marov, Y. A., Toon, O. B., & Turco, R. P. (1983). The clouds and hazes of Venus. In D. M. Hunten, L. Colin, T. M. Donahue, & V. I. Moroz (Eds.), *Venus* (pp. 484–564). Tucson: University of Arizona Press.
11. James, E. P., Toon, O. B., & Schubert, G. (1997). A numerical microphysical model of the condensational Venus cloud. *Icarus, 129*, 147–171.
12. Kanas, N., & Manzey, D. (2008). *Space Psychology and Psychiatry* (2nd ed.). El Segundo: Microcosm Press and Dordrecht: Springer.

13. Bouwmeester, D., Pan, J.-W., Mattle, K., Eibl, M., Weinfurter, H., & Zeilinger, A. (1997). Experimental quantum teleportation. *Nature, 390,* 575–579.

14. Houtkooper, J. M. (2006) *Retrocausation or extant indefinite reality? Frontiers of Time: Retrocausation – Experiment and Theory* (edited by D.P. Sheehan). 87th Annual Meeting of the AAAS Pacific Division (20–22 June 2006). San Diego, CA (USA), Conference Proceedings Volume 863, 147–168.

15. Klein, H. P. (1999). Did Viking discover life on Mars? *Origins of Life and Evolution of Biosphere, 29,* 625–631.

16. Levin, G. V. (2007). Possible evidence for panspermia: the labeled release experiment. *International Journal of Astrobiology, 6,* 95–108.

17. Houtkooper, J. M. & Schulze-Makuch, D. (2007). A possible biogenic origin for hydrogen peroxide on Mars: the Viking results reinterpreted. *International Journal of Astrobiology, 6*(2), 147–152.

18. Navarro-González, R., Vargas, E., de la Rosa, J. Raga, A. C., & McKay, C. P. (2010). Reanalysis of the Viking results suggests perchlorate and organics at midlatitudes on Mars. *JGR-Planets* 115. doi: 10.1029/2010JE003599.

19. McKay, C. P., & Davis, W. (2013). Astrobiology. In *Encyclopedia of the Solar System* (3rd ed.). Waltham: Academic Press. In press.

20. Schulze-Makuch, D. (2002). *Evidence for the discharge of hydrothermal water into Lake Lucero, White Sands National Monument, Southern New Mexico.* New Mexico Geological Society Guidebook, 53rd Field Conference, Geology of White Sands, p. 325–329.

21. Laval, B., Cady, S. L., Pollack, J. C., McKay, C. P., Bird, J. S., Grotzinger, J. P., Ford, D. C., & Bohm, H. R. (2000). Modern freshwater microbialite analogues for ancient dendritic reef structures *Nature, 407,* 626–629.

22. Schulze-Makuch, D., Lim, D. S. S., Laval, B., Turse, C., Antonio, M. R. S., Chan, O., Pointing, S. B., Reid, D., & Irwin, L. N. (2013). Pavilion Lake microbialites: morphological, molecular, and biochemical evidence for a cold-water transition to colonial aggregates. *Life, 3,* 21–37.

23. Shapiro, R. S., & Schulze-Makuch, D. (2009). The search for alien life in our solar system: Strategies and priorities. *Astrobiology, 9,* 335–343.

24. Hueso, R., & Sanchez-Lavega, A. (2006). Methane storms on Saturn's moon Titan. *Nature, 442,* 428–431.

25. Tokano, T., McKay, C. P., Neubauer, F. M., Atreya, S. K., Ferri, F., Fulchignoni, M., & Niemann, H. B. (2006). Methane drizzle on Titan. *Nature, 442,* 432–435.

26. Stofan, E. R., Elachi, C., Lunine, J. I., Lorenz, R. D., Stiles, B., Mitchell, K. L., Ostro, S., Soderblom, L., Wood, C., Zebker, H., Wall, S., Janssen, M., Kirk, R., Lopes, R., Paganelli, F., Radebaugh, J., Wye, L., Anderson, Y., Allison, M., Boehmer, R., Callahan, P., Encrenaz, P., Flamini, E., Francescetti, G., Gim, Y., Hamilton, G., Hensley, S., Johnson, W. T. K.,

Kelleher, K., Muhleman, D., Paillou, P., Picardi, G., Posa, F., Roth, L., Seu, R., Shaffer, S., Vetrella, S., & West, R. (2007). The lakes of Titan. *Nature, 445,* 61–64.

27. Schulze-Makuch, D., & Grinspoon, D. H. (2005). Biologically enhanced energy and carbon cycling on Titan? *Astrobiology, 5,* 560–567.

28. McKay, C. P., & Smith, H. D. (2005). Possibilities for methanogenic life in liquid methane on the surface of Titan. *Icarus, 178,* 274–276.

29. Baross, J. A., Benner, S. A., Cody, G. D., Copley, S. D., Pace, N. R., Scott, J. H., Shapiro, R., Sogin, M. L., Stein, J. L., Summons, R., & Szostak, J. W. (2007). *The Limits of Organic Life in Planetary Systems.* Washington DC: National Academies Press.

30. Fortes, A. D. (2000). Exobiological implications of a possible ammonia-water ocean inside Titan. *Icarus, 146,* 444–452.

31. Schulze-Makuch, D., & Irwin, L. N. (2008). *Life in the universe: Expectations and constraints* (2nd ed.). Berlin: Springer.

32. Benner, S. A., Ricardo, A., & Carrigan, M. A. (2004). Is there a common chemical model for life in the universe? *Current Opinion in Chemical Biology, 8,* 672–689.

33. Wilson, E. (1980). *Sociobiology.* Cambridge: Harvard University Press.

34. Mueller, U. G., Rehner, S. A., & Schultz, T. R. (1998). The evolution of agriculture in ants. *Science, 281,* 2034–2038.

35. Brady, S. G. (2003). Evolution of the army ant syndrome: The origin and long-term evolutionary stasis of a complex of behavioral and reproductive adaptations. *Proceedings of the National Academy of Sciences of the USA, 100,* 6575–6579.

36. Irwin, L. N., & Schulze-Makuch, D. (2011). *Cosmic biology: How life could evolve on other worlds.* Omaha: Praxis Publishing Ltd.

37. Bechtel, W., & Richardson, R. C. (1998). Vitalism. In E. Craig (Ed.), *Routledge encyclopedia of philosophy.* London: Routledge.

38. Webb, S. (2002). *If the universe is teaming with aliens… Where is everybody? – Fifty solutions to the Fermi paradox.* Omaha: Praxis Publishing Ltd.

39. Wallenhorst, S. G. (1981). The Drake Equation re-examined. *Quarterly Journal of the Royal Astronomical Society, 22,* 380–387.

40. Von Neumann, J. (1966). *Theory of self-reproducing automata* (edited and completed by A. W. Burks). Champaign: University of Illinois Press.

41. Bostrum, N. (2003). Are we living in a computer simulation? *The Philosophical Quarterly, 53,* 243–255.